Taking Nazi Technology

Taking Nazi Technology

Allied Exploitation of German Science after the Second World War

DOUGLAS M. O'REAGAN

Johns Hopkins University Press

BALTIMORE

Johns Hopkins University Press
2715 North Charles Street
Baltimore, Maryland 21218-4363
www.press.jhu.edu

Library of Congress Cataloging-in-Publication Data

Names: O'Reagan, Douglas, 1985– author.
Title: Taking Nazi technology : Allied exploitation of German science after the
Second World War / Douglas M. O'Reagan.
Description: Baltimore : Johns Hopkins University Press, 2019. | Includes
bibliographical references and index.
Identifiers: LCCN 2018036810 | ISBN 9781421428871 (hardcover : alk. paper) |
ISBN 9781421428888 (electronic) | ISBN 1421428873 (hardcover : alk. paper) |
ISBN 1421428881 (electronic)
Subjects: LCSH: Technology transfer—Germany—History—20th century. | World
War, 1939–1945—Reparations. | World War, 1939–1945—Science. | Military
research—Germany—History—20th century.
Classification: LCC T174.3 .O74 2019 | DDC 338.943/0609044—dc23
LC record available at https://lccn.loc.gov/2018036810

A catalog record for this book is available from the British Library.

*Special discounts are available for bulk purchases of this book. For more information,
please contact Special Sales at 410-516-6936 or specialsales@press.jhu.edu.*

Johns Hopkins University Press uses environmentally friendly book materials,
including recycled text paper that is composed of at least 30 percent post-consumer
waste, whenever possible.

contents

acknowledgments

Serious historical research would be impossible without the work of countless archivists and librarians, some of whom I had the pleasure of working with directly but many more of whom contributed enormously to this work indirectly. Thanks to the staff at the US National Archives in College Park, Maryland, and Washington, DC; the Library of Congress; the US Army Center of Military History at Fort McNair; the Hagley Museum; the Science Heritage Institute; the UK National Archives in Kew; the British Library; the Archives Nationales in Fontainebleau and Paris; the Service Historique de la Defense; and the Bundesarchiv Koblenz and Bundesarchiv Berlin. This research would also have been impossible without the efforts of the librarians and library staff at University of California, Berkeley, the Berkeley Public Library, Stanford University, the University of Virginia, Washington State University, the Massachusetts Institute of Technology, and elsewhere.

I have been supported by generous fellowships, grants, and stipends from a number of groups, without which none of this would be possible. Thanks to UC Berkeley's Department of History, Graduate Division, Center for German and European Studies, and Institute for European Studies. Thanks to the US Department of Energy; the Consortium for the History of Science, Technology and Medicine; the Science History Institute; and the Miller Center for Public Affairs at the University of Virginia. Thanks also to those for whom I have worked while writing this book: the Fung Institute for Engineering Leadership in UC Berkeley's College of Engineering, Washington State University Tri-Cities and its Hanford History Project, and MIT. Each of these opportunities—and especially the relationships that came with them—has been fundamental in writing this book. Thanks specifically to Mabel Lee, Lee Fleming, Guan-Cheng Li, Michael Mays, Robert Franklin, Jillian Gardner-

Andrews, Jeffrey Ravel, Mabel Chin Sorett, and others who made those roles such exciting adventures.

I cannot hope to name all of the historians and other researchers whose comments, guidance, writing, and other contributions fed into this book, but I will mention at least a few (in addition to those listed in the bibliography). Thanks to Cathryn Carson and other UC Berkeley faculty who guided me during my pursuit of my PhD, the research from which was the starting point for this book. Hilary Falb Kalisman has been particularly generous with her time and feedback. The book benefited enormously from detailed comments by Ashley Leyba, James Hershberg, John Brown, Daniel Sargent, John Connelly, and Kristie Macrakis, as well as from the feedback from anonymous reviewers and the articles that fed into it. Thanks to the Johns Hopkins University Press team, including, but certainly not limited to, Matt McAdam and copyeditor Nicole Wayland.

Finally, I am grateful to my family and friends, and above all, thanks to Kaileigh Callender. I could not have made it through this project without your support.

AACP	Anglo-American Council on Productivity
ADI	American Documentation Institute
AEG	Allgemeine Elektricitäts-Gesellschaft
AMS	Aslib Microfilm Service
Aslib	Association of Special Libraries and Information Bureaux
BASF	Badische Anilin- und Soda Fabrik
BIOS	British Intelligence Objectives Subcommittee
CCG/BE	Control Commission for Germany, British Element
CCSDN	Comité de coordination scientifique de la defense nationale
CIA	Central Intelligence Agency
CIOS	Combined Intelligence Objectives Subcommittee
CNRS	Centre national de la recherche scientifique
CoCOM	Coordinating Committee for Multilateral Export Controls
DSIR	Department of Scientific and Industrial Research
EPES	Enemy Personnel Exploitation Section
FIAT	Field Information Agency, Technical
FID	Federation for Information and Documentation
G-2	intelligence division
GDR	German Democratic Republic (East Germany)
GPU	State Political Directorate (Gosudarstvennoye politicheskoye upravlenie)
JCS	Joint Chiefs of Staff
JIOA	Joint Intelligence Objectives Agency
KGB	Committee for State Security (Komitet gosudarstvennoy bezopasnosti)
KWG	Kaiser Wilhelm Society (Kaiser Wilhelm Gesellschaft)

MPG	Max Planck Gesellschaft / Max Planck Society
NKVD	People's Commissariat for Internal Affairs (Narodnyy Komissariat Vnutrennikh Del)
NRC	National Research Council
NTO	Science and Technology Offices (Nauchno-teknicheskii otdel)
OMGUS	Office of the Military Government, United States
OSRD	Office of Scientific Research and Development
OSS	Office of Strategic Services
OTS	Office of Technical Services
SAG	Sowjetische Aktiengesellschaft
SHAEF	Supreme Headquarters, Allied Expeditionary Force
SIAS	Scientific Intelligence Advisory Section
SOPRODOC	Société de productions documentaires
SVAG	Soviet Military Government in Germany
TIIC	Technical Industrial Intelligence Committee (at some points known as the Technical Industrial Intelligence Division [TIID])
VINITI	All-Russian Institute for Scientific and Technical Information

Taking Nazi Technology

Introduction

German science and technology was a terrifying threat to the Allied nations during the Second World War.[1] The first long-range, self-propelled missile, the V-2, shattered homes and terrified the citizens of London. Combined with Germany's generations-old reputation for excellence in science and engineering, the V-2 and other weapons gave credence to Nazi propaganda about forthcoming "wonder-weapons" that would turn the war decisively in the Axis' favor. The First World War's advances in poison gases, explosives, and machine guns had devastating effects, resulting in the death of about one out of every twenty individuals living in Axis countries and one out of one hundred in Allied nations. The new weapons of the Second World War—radar, rockets, better tanks and submarines, and, at the very end of the war, jet airplanes and atomic bombs—were proving equally decisive. Allied military leaders were left to wonder what else was in store.

Throughout the war, Allied intelligence agencies worked hard at uncovering the secrets of German military technology. As the German army fell farther back into Germany itself, teams of investigators raced along the front lines, seizing production and design facilities. They hoped to learn what weapons the Nazis might have passed on to Japan, and to redirect as many of these German advances as possible toward speeding up the conclusion of the Pacific War. Especially intriguing were tank designs, prototypes of experimental aircraft, any signs of progress on a German atomic bomb, and, of course, the V-2 missile.

None of the conquering armies stopped at just searching for military technologies, though. Nor did they cease their investigations after Japan surrendered in September 1945. Quite the opposite, as American, British, French, and Soviet forces occupied Germany, they orchestrated the largest-scale technology transfer program in history, aimed at almost every field of industrial technology and academic science. Swarms of investigators recruited from industry, military branches, and intelligence agencies scoured Germany's factories and research institutions. They seized or copied all kinds of documents, from patent applications to factory production data to science journals. They questioned, hired, and sometimes even kidnapped hundreds of scientists, engineers, and other technical personnel. They studied technologies from aeronautics to audiotapes, toy making to machine tools, chemicals to carpentry equipment. They grabbed academic libraries, jealously competed over chemists, and schemed to deny the fruits of German invention to any other nation—including their allies.

This book is a comparative history of the American, French, British, and Soviet efforts to transfer German science and technology to their own industries, academic research facilities, and military arsenals during the post–Second World War occupation of Germany. When I began studying this topic, I wanted to uncover why each of these nations would pursue these "intellectual reparations" in such similar ways and on such a scale, despite their very different economic, political, and diplomatic positions coming out of the war.[2] The Western Allies (the United States, United Kingdom, and France) both cooperated and competed in their efforts. The Soviet Union remained largely on the outside, attempting its own reparations program aimed at recovering from a far greater degree of devastation and preparing for expected conflict with the capitalist West. I found major differences in how each nation pursued German science and technology based in large part on what policymakers in that country saw as their biggest threats and opportunities in the emerging postwar world.

Plans for capturing this craft knowledge varied even within each nation, with many decision-makers pushing diverse schemes. Each nation deployed a tangle of independent or semi-independent entities to Germany, hoping to acquire different slices of German science, and often internal competition was as fierce as international competition. *America* did not have a plan for taking German science and technology, but lots of American military departments, civilian groups, intelligence agencies, and individual policymakers and businessmen had plans. Each of these nations (and the bureaucratic actors within them) also changed their overall priorities as the occupation went on and as internal politics met the international diplomacy of the early Cold War. I attempt to capture some of this fundamental messiness throughout the book. Decision-makers both at home and in Germany were operating with limited information, navigating complex and evolving bureaucracies, and sometimes acted with more eagerness and ambition to take German technology than foresight and deep planning.

Yet, as I researched, I kept coming across a debate that seemed to be going on in all of these nations at once and which was much more fundamental to scientific espionage: what does it take to transfer technology, anyway? Is it enough to copy documents? That would be convenient, especially with the help of the breakthrough information technology of the day: microfilm. With an army at their backs, investigators could copy any documents they wanted, including secret blueprints, raw experimental data, patents, and even academic journals. Microfilmed reports could be reproduced and sent around the world efficiently. Many of these exploitation programs depended on this premise.

As these investigations stretched over months and years, however, businessmen, politicians, and military researchers around the world came to a similar conclusion: technology lived at least as much in people as it did in things. All the blueprints and prototypes in the world could not capture the hands-on experimentation and skill—the "know-how," as they called it—that was absolutely necessary in order to really use and understand any kind of technology. Copying documents was useful, but no reports, however well written, would ever be enough. They needed the know-how. That meant acquiring German scientists and technical personnel, through hiring or even by force, even though anti-German sentiment made this politically tricky. It also meant sending their own engineers over to Germany for long-enough periods to acquire the skills through sustained, hands-on tutelage.

This book falls into any number of historical fields: the history of science and technology; diplomatic history; business history; the history of ideas; the history of Germany, the United States, France, the United Kingdom, and the Soviet Union; the history of espionage and intelligence; and potentially others. It is, among other things, a history of the diplomatic and political consequences of cultural ideas about technology and society. This was an era when science and technology became increasingly important for national security and increasingly important trading chips among nations. Diplomacy, domestic science policy, plans for rebuilding Germany, industrial policy—all became deeply entangled in the early postwar years. Taking German technology mattered for more than just a few industries and went well beyond the famous cases of V-2 rockets, unethical medical experiments, and the questionable decision to let some "Nazi" scientists off the hook in exchange for service. The legacies of these exploitation efforts reverberated in intellectual property law and policy, international business tactics, diplomatic relations, and how states promoted industrial technology for decades to come.

In addition to academic audiences, I wrote the book in part for a general audience without an exhaustive knowledge of the historical context surrounding these programs. I have attempted to include enough detail and explanation that readers should not need more than a basic knowledge of the end of the Second World War and the early Cold War to follow along. With the exception of a few cases where non-English words look and translate almost exactly into English (e.g., the French *recherche scientifique* for "scientific research"), I have included my own loose translations, in notes if not in the text itself. An interest in science and technology will certainly make things more interesting, but detailed knowledge of these fields should not be needed.

Reaching different audiences means that some sections will not appeal to everyone equally. The average reader stands to gain more from chapter 5, for example, than will specialists in Soviet history who are already familiar with that story. Some subsections address issues of special interest to historians of technology, business, intelligence, diplomacy, and other fields. There is some basic repetition throughout, so even if readers skim some segments, the rest of the book should hold together. Details matter, but in a complicated, fascinating story like this one, it is far more important that everyone gets a chance to see the forest than that each person study every tree.

Historical Context

Entering the Second World War, Germany had a centuries-old reputation for scientific and technological leadership. In mathematics, Germany overtook France as the world's leader by the mid-nineteenth century.[3] From the Nobel Prizes' founding in 1901 through the late 1930s, German scientists won fourteen prizes in Chemistry, eleven in Physics, and nine in Physiology or Medicine. German scientists (by citizenship) earned more Nobel Prizes than any other nation, from the Prizes' founding through 1956 (barring one year when it tied with the United Kingdom in 1904–1905), after which the United States took over. Counting solely science prizes, Germany led until 1964.[4]

Even after the damage wrought by the First World War, Germany remained in a central position.[5] Throughout the late nineteenth and early twentieth centuries, many of the best science students from around the world took pilgrimages to train in German laboratories and universities. Among them were American scientists J. Robert Oppenheimer and Irving Langmuir, and many who would eventually immigrate to America, such as Eugene Wigner (born in Hungary) and Enrico Fermi (Italy). Likewise, nearly nine thousand British students studied in German universities between 1849 and 1914.[6] By the 1940s, the United States had overtaken Germany in a number of metrics (e.g., number of articles published or number of PhDs), and American PhD programs rose in prestige to rival the Germans', but Germany's overall reputation survived.[7]

British admiration for German science tied into a broader admiration for German learning and culture, as well as a long-standing fear of German military aggression.[8] In the late nineteenth century, a steady flow of German scientists found jobs in the United Kingdom, with major benefit to British science and industry.[9] The founding of the journal *Nature* in 1859 gave the British additional claim to international prominence in science, competing with a traditional German strength in science publishing. However, now that *Nature* could provide a single (English-language) review of the international scene, British readers became all the more aware of Germany's scientific strength. As historian Rainald von Gizycki put it: "Its British readers obtained from it a clearer picture of the centrality of German science. . . . *Nature* returned incessantly to this theme with the gravity and insistence of Cato warning Rome of the danger of Carthage."[10]

Close ties between universities and industrial research facilities were one key to German scientific strength, with the chemical industry as a major benefactor and beneficiary.[11] The chemical industry was one of the largest in the world in the first half of the twentieth century, as fertilizers, dyes, and explosives all gained economic importance. German cartels held a dominant market share in the international chemical trade. They bought out chemical companies in other countries and used intellectual property (patents and trademarks) to control others. The Reichspatentamt, or State Patent Office, had an international reputation for efficiency, and policymakers around the world used it as a model when considering how to reform their own systems.

During the First World War, this strength in the chemical industry was a major advantage for Germany. The war has been dubbed "the chemists' war" because of the impact of explosives, poison gases, pharmaceuticals, and fuels. In each area, the Triple Alliance of Germany, Austria-Hungary, and Italy had an advantage over the British, French, Russian, and, eventually, American forces.[12] In response, both the American and British governments seized German patents (at first with a nominal goal of keeping them in neutral custody throughout the war, then later selling them off) under the Trading with the Enemy Acts.[13] In America, the Office of the Alien Property Custodian sold all chemical patents to the Chemical Foundation, an industry group organized for this purpose, who in turn licensed the patents to American chemical firms for very little money. After the war, the German cartels sued the Chemical Foundation in US courts for return of their intellectual property and for damages but lost, as the foundation was able to argue successfully that the patents alone were of little value, since they did not have enough information to transfer the technology themselves. These patents became something like "intellectual reparations," though on a relatively small scale and not planned that way initially.

German Leadership: Perception and Reality

In retrospect, Germany's lead in science and technology—such that it existed, in fact, in the first place, outside of perceptions—eroded considerably in the first half of the twentieth century. As Volker Berghahn has argued: "Neither the pre-1945 relationship between [the US and Germany] nor what happened afterwards can be understood without conceiving of the role of technology in modern industrial societies in much broader terms than patents and machines."[14] Including "technologies" such as industrial organization and business management techniques, you can tell a convincing and important version of this history in which Germany looked to America for leadership as early as the late nineteenth century.

American industry jumped in rankings of gross domestic product from the 1860s (i.e., the end of the American Civil War) through the early twentieth century, and these advances certainly led governments and businessmen around the world to reflect on what they could learn from American practices. "Scientific management," or Taylorism (after its inventor, Frederick Taylor), is one example of American leadership in industrial organization. From its origins in the 1880s, it became an international phenomenon in the 1900s to 1920s, leading business magnates from as far away as the Soviet Union to visit and study American factories.[15] In this broader context of industrial organization and business technology, American leadership (including over Germany) began long before the post–Second World War boom.[16]

In the more specific world of academic research, the Nazi party's racism and antisemitism undercut German science to a tremendous degree.[17] The party's rise drove many of Germany's brightest scientific and technical innovators to flee abroad, and anti-Jewish laws forced the departure of many of those who had not left willingly. Hundreds of scientists were among those who fled in the 1920s and 1930s, most of whom ended up in America, Britain, and France. These emigrant scientists included Albert Einstein, Max Born, Hans Bethe, and Karl Popper. At least fifteen of these scientists would go on to win Nobel Prizes in scientific fields. This mass exodus seriously damaged German science (and in fields well beyond science, as many great artists and other scholars fled) while strengthening the countries who took these individuals in. Several of these émigrés eventually played important roles in the Manhattan Project, for example.[18]

Both world wars led to sharp anti-German sentiment in America, Britain, France, and Russia/the Soviet Union, and German scientists found themselves excluded from international organizations for much of the 1920s through the 1940s. Given the growing importance of these other nations in the science publishing marketplace and in hosting conferences, this exclusion dealt another major blow to German science.[19]

These developments are likely much clearer in hindsight than at the time, and Germany retained much of its image of technological sophistication as the Second World War approached. Future chapters include additional illustrations of German technology's reputation. The Nazi party cultivated an image of modernity, employed cutting-edge media and propaganda technologies, drew on ties between Italian fascism and futurism, and spoke continually about a society based on race science (at the same time, they disparaged modern European society and rationalism).[20] During the war itself, Germany developed or deployed several innovative military technologies, including jet engine aircraft (albeit at the very end of the war), advances in small arms (particularly high-quality tanks), and most famously, the V-1 flying bomb and the V-2 missile.[21]

Given this background, it is perhaps unsurprising that Allied occupiers would investigate and try to learn from German science and technology. Editorials in trade journals, newspaper op-eds, and government planning documents all promoted the idea. The idea was clearly appealing to more than just a few policymakers. Sidney Kirkpatrick, editor of *Chemical and Metallurgical Engineering*, was a major promoter of such efforts, writing in May 1945: "We are not seeking reparations in money, goods, or land. But what we can obtain in the way of new science and technology, processes, patents, and know-how can be used by the democracies in building a better and safer world."[22] An engineer lamented in *Aero Digest* in April 1946 that "We'll Just

Never Learn:... Very few persons realize that the technical information which we have 'liberated' in Germany is one of the biggest 'reparations' we may ever receive."[23] When officials in the United Kingdom offered to include its empire in British efforts, representatives from Canada, Australia, India, and South Africa eagerly accepted (and, in fact, South Africa signed on in 1948 after having tried to operate such a program independently).[24] As I discuss in later chapters, the reactions to German science and technology varied considerably across industry, but the initial interest was widespread and intense.

Historiography

These scientific exploitation programs became public knowledge even while they were still under way, and ever since, the brilliant but amoral Nazi scientist working for other nations has been a fixture of popular culture. The most prominent real-life scientist of this type was Wernher von Braun, one of the chief designers of the V-2 rockets and later an important team leader in the American space program.[25] Von Braun became an influential promoter of science to the public, even teaming up with Walt Disney to produce "edu-tainment" films about space exploration in the 1950s. He was one inspiration for famous fictional scientists, from the title character in *Dr. Strangelove* to the villain in the 2014 film *Captain America: The Winter Soldier*.

Every few decades since the war, another journalist or popular historian has written an expose about American use of former Nazi scientists, especially as it relates to Operation Paperclip.[26] From Michel Bar-Zohar's *The Hunt for German Scientists, 1944–60* (1967) to Clarence Lasby's *Project Paperclip* (1971), Tom Bower's *The Paperclip Conspiracy* (1987), Linda Hunt's *Secret Agenda* (1991), and Annie Jacobsen's *Operation Paperclip* (2014), among others, the topic continues to interest new audiences. Historian Brian Crim added to this collection with a well-researched, more academic *Our Germans* in 2018. The quality of these works varies tremendously. Jacobsen's and Crim's are the best researched, benefiting from much easier access to historical records. They almost invariably aim to shock audiences with the moral outrage of employing former Nazis and in a few cases draw clear lines between war crimes in Germany, American policymakers' knowledge of those crimes, and the scientists being brought to work in the United States regardless (and against immigration laws). As this book shows, Operation Paperclip was actually just one small part of a huge constellation of related programs, within and far beyond American shores, but sometimes a narrow focus can make for more compelling and useful stories. The moral questions these exposés explore are important, but they are not the focus of this book.

Among academic historians, the exploitation of German science has received much less attention, though there are some important exceptions. The foremost among them is John Gimbel's *Science, Technology, and Reparations*, published in 1990.[27] This book shook the episode from relative obscurity through a forceful and well-researched argument that this form of "intellectual reparations" benefited the United States to the tune of billions of dollars, a figure on par with what the Soviet Union seized in material reparations. Gimbel argued that although American diplomats argued for moral authority in the Cold War by claiming to have refused reparations, in contrast to the Soviets looting East Germany, this was actually disingenuous. In Gimbel's words: "I admit quite frankly that I am no closer to a precise evaluation [of the value of these 'reparations'] than anyone else. What I have been able to show, however . . . is that the amount and the value are by no means insignificant. The $10 billion figure bandied about by the Russians and their friends and dismissed by State Department functionaries as 'fantastic' is probably not far from the mark."[28]

Science, Technology, and Reparations set off a flurry of follow-up studies by historians in the United States, Germany, and, in rarer cases, in other countries. A conference in 1996 brought together many of these responses, collected in an edited volume titled *Technology Transfer out of Germany after 1945*.[29] The ten essays in this collection address a wide range of topics, including the influence the Nazi party had on German science, the idea of "intellectual reparations" within international law, American-German business relations during the twentieth century, and the importance of intellectual reparations within chemical and aeronautics industries. I address the specifics of several of these essays throughout the book. Of particular interest is Raymond Stokes's contribution on the chemical industry, in which he argues: "Neither Allied hopes nor German fears of the forced technology transfer programs were realized, although one of the programs' unintended effects was to promote integration of each of the postwar German states into the sphere of influence dominated by one of the two superpowers."[30] This is an argument with which I am in broad agreement. As I argue throughout the book, it applies considerably beyond the chemical industry—in fact, it is perhaps less true for chemicals than it is for many other fields—and runs contrary to much of the other writing on the topic, which tends to adopt Gimbel's research questions and conclusions.

Almost all of this writing and that not yet mentioned focus exclusively (or at least primarily) on the American experience, yet as I show in this book, this was far from solely an American enterprise. Some writing does exist on each of the other major Allied powers, some of it excellent, though there is substantially less. While I discuss these in more detail in each of the first four chapters, a few trends are worth noting at the outset. The French case has received the least study, though French historian Corine Defrance has written several articles on the topic to supplement broader histories of Franco-German relations.[31] Jacques Villain, a rocket scientist–cum-historian, has authored several related articles.[32] Most research useful for understanding French policy comes from historians studying broader trends, however: the development of the European Economic Community (and early European Union), diplomatic positioning after the war, and the importance of science for French national self-image.[33]

The most impressive study of British scientific exploitation policy is undoubtedly a 1,265-page dissertation by Carl Glatt.[34] A handful of articles attempt more accessible treatments of the subject.[35] Those seeking something in between, though, would do well to look in the relatively well-developed historiography of British espionage, paired with histories of British concern about retaining a prestigious role in the postwar world.[36]

Finally, while Soviet records are notoriously difficult to access, a number of excellent histories have built on East German archives, memoirs of German scientists taken to the Soviet Union, and Western intelligence assessments of Soviet activities.[37]

Consequences and Importance of Scientific Exploitation

Almost none of the aforementioned works tackle the exploitation of German science beyond a single nation's experiences. In contrast, this book is fundamentally a comparative history of each of the major Allied powers. This has several important advantages. It allows better understanding of the events themselves, which played significant and underappreciated roles in shaping the economies, diplomacy, and internal politics of the involved nations. In a sense, there was a sort of natural experiment at play: given different starting conditions, but similar objectives, how does scientific espionage play out on a similar playing field?

When studying just one nation's experiences, it is difficult to see whether something succeeded (or failed) because of something particular to that nation or because of a broader, international trend. There were a number of major changes in science's role in society under way in this period that other historians have already identified. Science grew exponentially during the twentieth century, for example, whether measured by the number of scientists, the number of journal articles, or several other metrics. Science and technology became important to diplomats and national security specialists in new ways. The comparative discussion here allows insight into how these broader trends played out in different contexts.

This book also moves away from one of the questions that has occupied nearly all of the academic histories related to this topic: the dollar-value worth of these "intellectual reparations." I do not attempt any sort of answer to that question besides the vaguest of discussion in the conclusion. The very question of how much these programs were *worth*, I argue, misses most of what makes these efforts important. Technology transfer is hard. What factors lead to successful and unsuccessful attempts to move technologies across national and cultural borders has been a driving question in the history of technology, and little consensus has ever been reached.[38] This book continues this line of questioning rather than assuming that technology transfer in general works *this* or *that* way and extrapolating from there the money taken from Germany or won by the occupiers.

Given four nations with a desire to learn from Germany's accomplishments and the power of an occupying army at their backs, as well as the differences between the occupying nations, what factors seem to have aided or impeded successful technology transfer and scientific communication? This is a central question examined in this book. In particular, I focus on how people at the time understood technology and what *they* thought it took to transfer it from place to place. As I argue, the 1940s was a period when this understanding changed, dramatically and importantly, toward a much heavier emphasis on the importance of "know-how."

The term "know-how," meaning something like "inventions, processes, formulas, designs, skilled manual methods, [and] preferred sequences of industrial operations learned from practical experience,"[39] saw a dramatic increase in usage starting in the 1940s. This attention paid to intangible knowledge, in turn, had serious implications for policy in Germany as well as for the wider postwar business and legal worlds. It was not the basic idea of know-how—or "tacit knowledge," as it is sometimes called—that was new. As businessmen and policymakers started to pay increased attention to this hands-on, craft knowledge, though, it meant changing the basic plans for taking German science.

Structure and Overview

This book begins with chapters covering each nation's story individually, then moves on to chapters addressing broader themes that tie these stories together. The first four chapters, then, cover the United States, United Kingdom, France, and the Soviet Union, in that order. These are not truly isolated histories, but treating them separately at first helps introduce the people and organizations involved without becoming overwhelming. The final three chapters then tackle how this history ties into larger historical forces that shaped and were shaped by the exploitation of German technology. This means that the book moves from a narrower to a wider lens.

Chapter 1 tells the most familiar story—the American exploitation of German technology—but I argue that both historians and journalists have misunderstood its importance. Rather than focus on the usual contrast between American values and the use of Nazi scientists, I emphasize in this chapter the contrast between American expectations about what they would find in Germany and what they actually did. Both the popular and academic histories about this topic focus on the dollar value of these reparations, yet this emphasis has problems in both theory and practice. American companies expressed great enthusiasm for investigations of German science and technology. However, once there, they frequently wrote about their disappointment with what they found. The greatest discovery of American industrial investigators in Germany was not a cache of new technologies or scientific secrets (though they found those, too) but rather a new perception that American technology and industry—not German—led the world.

In chapter 2, I turn to British efforts, as they sought to balance two increasingly urgent priorities: close relations with the United States and developing more export industries. The Americans seemed likely to be a keystone of British postwar security, but export industries were the only way to pay down national debt and repair balance-of-trade issues. At first, these seemed to be harmonious goals, as Anglo-American cooperation seemed sure to help both nations. Amid parliamentary debates about the relationship between scientific knowledge and gaining economically useful technology, British industrial investigators scoured Germany for patents, blueprints, trade secrets, and, in some cases, skilled personnel. However, initial plans to use this cooperation to draw together American and British intelligence ran into the problem of how to copy German technology. As they lost faith in capturing this technology in written reports (that could be shared with their American allies), British policymakers turned toward a more self-interested exploitation style emphasizing "know-how."

In chapter 3, I address the French occupation of Germany and how French concerns with diplomatic standing in the world shaped domestic science policy and occupation policy in Germany. Unlike the close Anglo-American relationship, the French remained relatively isolated from its allies. This, in turn, meant that French policymakers had less reason to mimic American planning for the sake of good relations. Combined with important policymakers' fundamentally different understanding of technology transfer, this led to French exploitation efforts that differed from the other Allies' in important ways. French belief that science and technology were fundamentally part of the society around them led important policymakers to see no value in removing German scientists from their original contexts. Instead, they focused on acquiring German expertise and technology through collaboration and surveillance rather than transplantation, for example, by building Franco-German research centers. This policy likely had significant impact in easing tension between these former enemies, paving a path for cooperation in the early formative era of a European economic community. Conversely, these different scientific exploitation strategies only generated more friction among the Western Allies.

In chapter 4, I wrap up the national summaries with the Soviet Union's actions in Germany. This includes the famous Operation Osoaviakhim, in which thousands of German scientists and technicians found themselves essentially kidnapped deep into the Soviet Union. There, they worked in relative comfort (but with no freedom to leave) for years. They contributed to Soviet science and technology, including research on the atomic bomb, before returning to East Germany. This chapter, unlike the others in the book, is more of a synthesis of other historians' work than a substantially new interpretation built from primary sources. The main reason for this is access: gaining entry to Russian archives relating to intelligence and national security is difficult, often impossible, especially for a foreigner. Still, other scholars have made good use of German sources on the Soviet occupation, and this chapter draws from my own research on how the Western Allies *interpreted* Soviet actions. Using these studies, we can piece together enough of the Soviet story to draw parallels to (and differences from) the American, British, and French stories.

Chapter 5 is the first thematic chapter, in which I look at how concern with rehabilitating Germany's academic science tied into each of the occupiers' own changing domestic science policies. In each case, policymakers struggled consciously with what science could contribute to a nation. Was it a force of democracy, inspiring a society full of openness and civic disagreement? A dangerous source of military strength? A venue for cultural diplomacy, allowing nations to influence one another regardless of the content of the science itself? What did it take to generate economically or militarily useful science? In this chapter, I argue that the United States was not the only country to turn discussions of German science into sources of soft power diplomacy in the 1940s and early 1950s; the United Kingdom, France, and the Soviet Union did the same. Though the details varied by country, each nation saw science as a powerful agent for building and exerting state power. This thinking and the planning for the occupation of Germany were powerful influences on each other.

In chapter 6, I connect a basic, practical question about the exploitation of German science—*How* did intelligence agencies plan on moving around all that copied German documentation?—to major changes under way in scientific and technical communication in this era. The first half of the twentieth century witnessed a vast, exponential growth in the amount of scientific information that the world's science libraries, universities, and corporate research facilities sought to manage. In response, ambitious scientists, librarians, and other thinkers began promoting utopian solutions based on cutting-edge information technologies (primarily, microfilm). The exploitation efforts in Germany, with the vast amount of information gathered, proved a testing ground for these "documentation movement" activists. The major problems surrounding the capture of intellectual reparations were in part a reflection of the failure of the documentarians' microfilm-and-bibliography-based schemes. The attempt itself had more lasting legacies. Investigations of German science led governments to invest in information technologies (including early mechanical sorting), especially for use in intelligence agencies. They also set the stage for major changes in library science and information science.

Chapter 7 takes the widest lens. In it, I look at some of the major legacies of the intellectual reparations programs in Germany. One of the biggest lessons, learned by businessmen and politicians around the world, was that moving technology from place to place was extremely difficult and often impossible without also accounting for the "know-how." That meant focusing on the people who developed and worked with the technology at least as much as documents or blueprints. This interest in know-how became an international phenomenon in the postwar years, changing the face of international business and challenging lawmakers to rethink how they could influence the world around them. At the same time, as technology took on new significance for national security, policymakers started thinking through how they could prevent the spread of certain technologies, whether by limiting the movement of skilled people with the know-how or by other means. While the exploitation of German technology was surely not solely responsible for these big-picture trends, it was at least one important contributor, and these long-term legacies are explored in this chapter.

This is a book about the different paths these four nations took in orchestrating possibly the most ambitious technology transfer programs ever attempted. Amid the drama of citizen-spies racing across battle lines and scientists enabling mass slaughter, policymakers around the world learned lessons that reshaped the postwar world. One of these lessons, and a key theme throughout the book, is that seizing technology is far easier said than done. These programs often failed or at least succeeded far more modestly than planners promised. Technologies exist to solve problems within particular societies, and so any technology transfer requires adaptation, trial and error, and sometimes the discovery that the source and recipient simply have different needs. In this case, German technology had quite a reputation, but in many cases the process of seizing German industrial science had such high costs that they likely erased any benefit. Technology transfer and scientific communication became ever more important in the postwar world, with a globalizing economy and Cold War espionage in the headlines. Even a simple lesson of being cautious about scientific espionage would have been invaluable.

Governments and businessmen also took other lessons from the intellectual reparations programs. The occupation zones were laboratories in which the Allied governments tested theories about how science and technology interacted with national security. As they simultaneously governed their occupation zones and prepared at home for the emerging Cold War, science policy became ever more important for a variety of policy goals—and thus so, too, did the programs designed to exploit German science. Science and technology sometimes seemed like very different things, and needed different policy prescriptions, but they also often ran together in this period.[40] The agents sent to copy Germany's science and technology did not draw sharp distinctions. Microfilm reels sent home from German research institutes often included a mix of patent filings, raw research data, papers from academic journals, notes from interviews with both professors and industrial engineers, and various other items.

This is also a book about how the efforts to untangle what science and technology meant for postwar Germany wrapped together with science policy on the home front. The lessons were not always clear-cut, nor the consequences direct, but they were far-reaching. The programs had legacies from scientific communication systems to industrial policy, from the development of espionage programs to international business law. One book cannot capture every legacy rippling out from complicated programs operating across many scientific fields, many industries, and spread over several continents. By bringing together as many of these as possible, though, I hope to show the breathtaking ambition and powerful impact of these programs, especially in how people around the world thought about moving technology from place to place. J. Robert Oppenheimer, "father of the atomic bomb," commented in an interview with *Time* magazine in 1948 that "the best way to send information is to wrap it up in a person."[41] However abstract this lesson might seem, it is one that pervades this book and reshaped the twentieth century.

1

American Exploitation Programs

High Hopes, Narrow Gains, and Long-Term Lessons

As the Allied Expeditionary Force marched east after D-Day, scientific intelligence units raced across the front lines securing buildings, equipment, and people deemed to have scientific or technical value. American and British troops escorted teams of scientists and moved technical documents to their own bases, especially from lands they would have to hand over to Soviet occupiers. After the war officially ended, with Germany divided into four zones of occupation, US planners established overlapping, sometimes conflicting agencies responsible for learning about German science and technology. These agencies convinced German scientists (sometimes forcefully) to migrate to America or Britain in order to both gain their skills and deny the same to the Soviets. Hundreds of investigators—mostly ordinary engineers borrowed from dozens of US firms—swarmed over Germany, seeking "intellectual reparations" in a wide variety of civilian industries. Nothing similar had ever been attempted on this scale.

This is an incredible episode in American history, but not just for the reasons that people often assume. Many books, movies, documentaries, and a few academic histories tell versions of this story. They tend to focus on the famous case of Operation Paperclip, which brought Wernher von Braun and his team of rocket scientists to America. Von Braun and his team became citizens and eventually National Aeronautics and Space Administration (NASA) employees, and they helped design the rockets that sent the first men to the moon. Von Braun was a canny self-promoter, and his celebrity has fueled the idea that the main story here is one of priceless German scientists at the center of an epic Cold War struggle. Comedian Bob Hope famously joked in the wake of the 1957 Sputnik launch: "All this goes to show that their Germans are better than our Germans," and he was far from the only person to think this way.[1] Yet even while some at the time asked why the United States had not done more to recruit German scientists, others pointed out the moral cost of co-opting *any* of this work: some Nazi science and technology, including its rocketry program, was built on inhumane experimentation, slave labor, and other war crimes. Was it acceptable to hire former Nazis and protect them from punishment if it meant a technological edge in the Cold War?

This question matters, but focusing on aeronautics and the worst imported war criminals warps our understanding of the full breadth and legacies of the American intellectual reparations programs. These programs were far broader than the famous cases of rockets and nuclear weapons. Teams of civilian and military investigators scoured German industrial science and technology in nearly every field imaginable, from synthetic oil production to wood pulping, toy manufacturing to machine tool construction, coal mining to building precision watches. Paperclip and Alsos, the mission to investigate how far the Nazi regime had gotten in nuclear weapon design, were just relatively small facets of this much bigger undertaking.

The extreme case of rocketry distracts from a fundamental question: Why would American policymakers and businessmen think they had so much to learn from Germany? American technology was no slouch going into the war. The American economy's gross domestic product was twice Britain and Germany's *put together* by 1900, and that distance only increased by 1940. Many German business magnates took trips to the United States in the early twentieth century to learn from American Taylorist efficiency, epitomized by Ford's Model T production line. War mobilization only increased this lead in productivity. Why, then, were so many American businessmen and politicians so certain they had so much to learn from German industry, across such a wide variety of fields?

Secondly, these intellectual reparations programs did not happen in isolation. They were part of the tense, fast-changing diplomacy of the wartime alliance and early Cold War. Even within the United States, politics of all kinds led to fractured, evolving overall policy for taking German science: infighting between the Democratic and Republican Parties, power struggles *within* those parties, the State Department at odds with military agencies, newly founded intelligence agencies vying for influence, and individual personalities clashing all played roles. This full context is needed in order to understand these programs, and these programs in turn offer a window into a turning point in US and world history, as world war settled into Cold War.

This chapter starts by looking back at the US-German relationship in science and technology before the war. That history helps us understand what American policymakers saw when they conceived of taking Nazi technology, though, as the following section shows, the actual path forward was winding. I hope readers come away with a sense of the massive scope of these programs and the intense interest from a wide variety of sources that made them a reality. The logistical hurdles were high, and only much smaller-scale programs would have been possible without buy-in across industries, with help from trade presses, various government agencies from Commerce to War, and university faculty and administrators across the country.

Despite that huge outpouring of support and excitement, I argue that the biggest lesson most American industrial investigators took away was that they had been deceived by Germany's reputation for scientific preeminence—and, as much, began planning for the postwar world under American leadership. To be sure, the United States gained extremely valuable information and expertise in some military technologies, and even a few civilian industries. I describe those gains in detail. In most cases, though, American expectations did not match reality, and this meant rethinking what American industry had to offer in the global economy of the postwar world. For decades—even centuries— American science had sent its brightest graduate students to study with German professors. American firms subscribed to German trade publications, sought licenses, and feared competition from German firms, and in some fields were stuck being subordinate to German cartels. In the postwar years, American leaders increasingly saw the United States at the forefront of international science and technology, and saw a new diplomatic lever in being able to offer American scientific and technical aid. In some ways, the most important story here is one of American industry taking on self-conscious leadership in the world, realizing they could be teachers rather than students, and forging productive business relationships with former mortal enemies. One counterintuitive consequence of this is that far from robbing West Germany, both nations might well have benefited from America's attempts at taking "the only reparations we are likely to receive."[2]

American and German Science and Industry before the Second World War

The history of relations between the United States and Germany in the realms of science and industry could fill several bookshelves, but at least a brief discussion is worthwhile here in order to grasp the mind-set of policymakers trying to decide how to handle postwar Germany.[3] In that sense, it is at least as important to assess *perceptions* as it is to assess *reality*. With the benefit of hindsight, the late nineteenth and early twentieth centuries saw the United States rise from being something of a backwater to an international powerhouse of science and industry, while Germany (and Britain and France, also traditional powers) grew more slowly and fell relatively behind. Still, big economic trends are not always immediately clear to those living through them, and reputations can outlast reality. The ongoing strength of Germany's high-technology industries (especially chemicals), combined with ongoing excellence in science, sustained its reputation throughout the Second World War. That, in turn, made Germany's industrial technology a tempting target.

Part of Germany's reputation for cutting-edge industrial science came from its long-standing excellence in academic science (which was, after all, connected through a variety of academic-industrial institutions). Throughout the nineteenth century, top-level American scientists-in-training traveled to Europe to pursue graduate studies, and especially to Germany. It is no coincidence that the first American to receive a Nobel Prize in Physics, Albert Michelson, studied at the Universities of Berlin and Heidelberg, where he worked with German scientists such as Hermann von Helmholtz. German-style graduate training slowly expanded in the United States in the late nineteenth century, starting from Johns Hopkins and reforms at Yale and Harvard. In the first half of the twentieth century, American science grew in size and stature. Still, Germany remained very strong in many fields. Between 1900 and 1941, sixteen German scientists received Nobel Prizes in Chemistry (compared to three Americans), and ten German physicists (versus six Americans). Nobel Prizes are hardly the be-all and end-all of scientific achievement but are at least a decent proxy for reputation. American policymakers in the 1940s would have grown up in a world with American science on the rise, but they would have been taught by those who held Germany in extremely high esteem.

In industrial settings, a healthy stream of engineers and managers traveled between the United States and Germany throughout the early twentieth century, investigating each other's production techniques, goods, and infrastructure. American innovations in standardized mass production, often grouped together under the terms "Taylorism" and "Fordism," became an international phenomenon in the first decades of the twentieth century, one sign of American industry's rise. German industrialists were among those eager to implement them. As one German engineering professor noted: "At first a few leading personalities came individually; then major firms sent their employees in groups of twos and threes. . . . Soon the passenger lists of the beautiful ships of Hamburg-America Line and of North German Lloyd looked like a register of the leading industrial firms of Germany."[4] Gustav Krupp, head of Krupp AG heavy industries, was only one of many leading industrialists who took such trips. The First World War broke long-standing connections between American and German firms, however, and this familiarity faded somewhat in the 1920s and 1930s.

Meanwhile, German cartels had international monopolies in one of the most visible, powerful, high-technology sectors of the international economy: chemicals. Articles in leading American newspapers throughout the 1930s reflected Germany's reputation for engineering and technological innovation, with titles such as: "Reports Germans Lead in Chemistry: Editor of Research Papers Says They Have Regained Pre-War Eminence" (1930), "Germany: Laboratory of the World" (1930), "Industrial Uptrend is Noted in Germany" (1935), "German Chemicals in Demand" (1937), and "US Held Enriched by German Exiles: Flight of Chemists is Called Boon to Science Here" (1939).[5] Many similar examples could be found in leading interwar American newspapers reflecting a mix of reality, fears, expectations, and, later, Nazi propaganda.

This final newspaper article points to another important episode that simultaneously showed off and (with benefit of hindsight) decimated Germany's scientific and technical prowess: the exodus of Jewish and other persecuted scientists from Nazi Germany in the 1930s. Faced with Hitler's rise and increasingly discriminatory laws, thousands of brilliant scientists, writers, artists, and intellectuals of all types fled Germany. Not all were allowed into America or Britain, in part due to antisemitism, but many intellectuals were able to use scholarly networks to find political support and employment needed for immigration.[6] By 1944, more than 133,000 German Jewish émigrés arrived in the United States, and among them were some of Germany's best minds.

Famous émigré physicists included Albert Einstein, Leo Szilard, Eugene Wigner, Edward Teller, John von Neumann, and Hans Bethe. Chemists include Otto Meyerhof, Otto Stern, Otto Loewi, Max Bergmann, Carl Neuberg, and Kasimir Fajans. A number of these scientists were key players in the Manhattan Project and other wartime research. Historian Reinhard Siegmund-Schultze has argued that this immigrant wave created new centers of mathematical excellence in the United States, reduced American mathematical provincialism, and dramatically increased the strength and standing of America's mathematical community overall.[7] Similar stories could be told in a number of fields. America's growing scientific capability actually worked against those trying to find employment for these refugees, as an editorial in the journal *Science* noted in 1940: "The advantages of scientific and technological superiority, once held by the Europeans, no longer exist and foreigners have not had the opportunities to make themselves useful that were enjoyed by refugees two and three generations ago."[8] Still, American science benefited enormously from this wave, once again reinforcing how much there was to gain from Germany.

Industrial science benefited, as did academic research. Among the German Jewish (and other) refugees who ended up in the United States were skilled craftsmen and industrial scientists, and they, too, sought to make themselves indispensable. Economist Petra Moser has studied the impact of the industrial chemists among these refugees and found that in areas of the chemical industry where these émigré scientists worked, invention and innovation (measured by patents) increased substantially.[9] Even after controlling for a number of factors you might consider (e.g., was patenting already increasing in these fields?), Moser estimates that in the specific subfields where German Jewish émigrés joined American industry, patenting increased by 31 percent more than in other subfields of chemistry. The exchange of ideas and movement of skilled workers generally does cause increased innovation, and in this case these industrial chemists were a shining example for industry of the potential value of workers who were experienced in German industrial labs—at least in this field.

Widespread belief in German scientific and technical superiority is understandable, then, but should not be mistaken for such a general, widespread lead still existing in truth. For one thing, German industry's leaders worried since at least the early twentieth century about the opposite trend: that America had passed them by. Around 1900, prominent German chemists worried that "German chemistry had been frequently surpassed by countries abroad," mostly meaning the United States. As a response, these chemists organized the Reich Chemical Association.[10] The Kaiser Wilhelm Society was founded in 1911 for just this reason—it was to spark greater academic-industrial collaboration and thus promote German technological leadership. The society was modeled in part on what its founders understood to be an American model, with science funding coming from semi-independent foundations and industry-academic pairings.[11]

Even Germany's apparent leadership in chemicals eroded in the interwar period, as politicians saw the importance of having local sources of explosives, pharmaceuticals, and dyes. American chemists who studied under German masters such as Adolf von Baeyer and Emil Fischer brought back new kinds of expertise, and American chemical firms reorganized during the First World War to assert some independence from the major German cartels. The wartime Trading with the Enemy Act even allowed the US government to seize German chemical patents, and then transfer those patents to the Chemical Foundation, an organization that would license them cheaply to US industry. With historical hindsight, it now seems that there was a fundamental shift toward American equality, if not leadership, in chemical industrial technology and productivity between the 1910s and 1930s.[12] American advances should not be overstated—Raymond Stokes convincingly argues that while "there may have been a relative decline in the absolute dominance of the cutting edge of the organic chemical industry by the German chemical producers in other countries . . . the Germans continued into the post-1945 period to be major players in international technological markets."[13] Still, the American chemical firms developed rapidly in the first half of the twentieth century, and perhaps even faster than they themselves realized relative to their German counterparts.

The idea that Nazi science and technology was uniformly cutting-edge and a major wartime advantage continues to live on, even in the work of many professional historians, though this is beginning to change. As historian Adam Tooze argues in his economic history of Nazi Germany, "it is hardly an exaggeration to say that historians of twentieth-century Germany share at least one common starting point: the assumption of a peculiar strength of the Germany economy."[14] Built on the prewar reputation of brands such as IG Farben, Krupp, Siemens, and Carl Zeiss, as well as wartime technologies that later became worldwide standards such as the V-1 and V-2 rockets, Panther and Tiger tanks, Me 262 jet fighter, and Mark XXI U-boat, the idea of German technological superiority was a "myth that appealed to numerous themes in postwar German political culture: regret at a chance of a victory wasted, the consolation provided by the supposed superiority of 'German technology,' the self-righteous commemoration of the horror of Allied bombing."[15] The rapid fall of France in 1940 gave a strong impression of technological superiority and innovative planning. More recently, military historians have argued that Germany started the war in 1939 with "no substantial technical superiority over the better-established military powers of the West" and blame the fall of France on a "fatal interlocking of Allied and German operational planning," but this was hardly obvious at the time.[16]

The aforementioned genuinely innovative wartime technologies (the V-2,

Mark XXI U-boat, etc.), though not developed quickly enough or brought into production on a sufficient scale to have a major impact on the war, were each part of wartime Nazi propaganda campaigns about "superweapons." As such, they, too, likely fed the perception that Germany was getting ahead. This assessment of potential Nazi superweapons seems less credulous when you consider that the United States and United Kingdom were both developing their own at that very moment: the atomic bomb. The Manhattan Project was born from a fear that Germany might get atomic weapons first, and one of the first scientific intelligence units sent to Europe with the invasion forces was the Alsos Mission to investigate the progress of a German atomic bomb.[17]

In the background of these shifts in relative industrial leadership, enormous economic and political crises raged: the First World War, the Great Depression, struggles over colonization and decolonization, and the buildup to the Second World War. Throughout, the United States lacked a permanent, civilian intelligence agency to provide officials with reliable, neutral information on conditions abroad. American policymakers and industrialists of this era hardly had ideal circumstances for a sober reassessment of how Germany's science and industry compared to American standards. As we will see, the persistent idea that German science was broadly superior fundamentally shaped the intellectual reparations programs and through them influenced the diplomacy of the early Cold War.

The American Investigation and Exploitation of German Science

During the early stages of the war, America had no systematic way of learning about German's scientific progress for a very simple reason: it had no real infrastructure for learning much of *anything* about foreign nations, beyond what policymakers themselves read in newspapers. For a variety of reasons (including a somewhat old-fashioned notion that "gentlemen do not open each other's mail"), America had never developed a permanent, civilian intelligence agency akin to Britain's Secret Intelligence Service (MI6), France's Deuxième Bureau, or the Soviet Union's Cheka/GPU/NKVD/KGB.[18] The branches of the American military had their own intelligence units prior to the war, but they were underfunded, interservice rivalry crippled their effectiveness, and the command structure did not invest in or trust wiretapping, code-breaking, or other "signals intelligence." As a result, information that could have warned about Pearl Harbor (had the messages been translated and analyzed) slipped through the cracks. The shock of Pearl Harbor left a lasting impression on policymakers and military leaders about the importance of gathering and processing intelligence.

A mandate and urgent desire for a capable intelligence community is a far cry from *having* an effective infrastructure, however, and much of the story of American intelligence, including scientific intelligence, in the 1940s is one of growing pains—amateurism, lax security standards, and redundancy—alongside some real successes. American scientific intelligence programs during the war must be understood in this context of rapid, sometimes haphazard growth and change. The Office of Strategic Services (OSS), predecessor to the Central Intelligence Agency (CIA), made some real wartime contributions but struggled with inexperienced, sometimes unprofessional employees—including double agents reporting to foreign powers. Military intelligence agencies reorganized to streamline analysis procedures but continued to suffer from interservice rivalry even under the newly formed Joint Chiefs of Staff. British counterparts provided some tutelage in running an intelligence service (one major source of the "special relationship" that exists between the countries today), but there was a great deal of work to be done in a short time, and British collaborators often felt that they gave more intelligence and aid than they received. Finally, in the context of total war and the shock of Pearl Harbor, creating redundant intelligence efforts often seemed simpler than deciding efficient lines of authority.

As a result, efforts at extracting German science and technology were a mess of overlapping jurisdictions, military and civilian organizations, and confusing acronyms and code names. There is no hope of a completely clear retelling of who did what and when. Even those involved at the time were often confused. This is true even when we isolate just the American programs, as we do in this chapter, though in reality the American, British, and to some degree the French efforts were deeply interwoven (and for that matter, more than a few American operatives were passing information to the Soviets). Both high-level diplomacy and on-the-ground decision-making in each nation influenced the others in important ways. Future chapters begin to unpack some of these connections.

This complexity is itself an important point. The growth so many agencies trying to accomplish the same goal—that is, to study and then acquire German industrial science—shows that the idea occurred and appealed to many different groups across a number of countries. It also helps explain some of the major differences we will see in how those involved described their successes and failures. For every generality (i.e., "these programs were often ineffective"), there are clear and important counterexamples. Before attempting to streamline and clarify, then, let us first embrace the bureaucratic mess. If nothing else, it might allow us to appreciate the complexity Germans faced in trying to decipher their postwar obligations to the occupiers and that policymakers faced in trying to coordinate and simplify the system. Remembering each of the following agencies is not crucial. I will reintroduce them when relevant and will otherwise refer to the broader efforts as "FIAT-related" or "FIAT-like" programs.

Acronym Soup: The Unruly, Shifting Mix of Investigative Efforts

During the war itself, the Supreme Headquarters, Allied Expeditionary Force (SHAEF) operated so-called T-Forces, groups of intelligence specialists, technicians, interrogators, translators, engineers, bomb squads, and combat troops drawn from Eisenhower's Allied Expeditionary Force. The primary duty of these T-Force units was to identify and secure intelligence targets in occupied territories, and this mandate extended to targets of primarily scientific and industrial value, not just those of purely military importance.[19] The T-Forces, though a unit of the Special Sections Sub-division of SHAEF's G-2 (intelligence) unit, were assigned targets by the Combined Intelligence Objectives Subcommittee (CIOS), a joint British–American task force. The task force was comprised of a British component, the British Intelligence Objectives Subcommittee (BIOS), and an American component, the Technical Industrial Intelligence Committee (TIIC) (at some points known as the Technical Industrial Intelligence Division [TIID]), the latter of which was established by the Joint Chiefs of Staff under the Joint Intelligence Committee. Represented on TIIC were the Foreign Economic Administration (FEA), Naval Intelligence, the Army G-2, the Army Air Staff, the Department of State, the OSS, the War Production Board, and the Office of Scientific Research and Development (OSRD).

Within the Special Sections Sub-division of SHAEF G-2, there also existed a Scientific Intelligence Advisory Section and the Enemy Personnel Exploitation Section, the latter of which was responsible for operating internment camps named "Dustbin" and "Ashcan," where scientific and technical personnel (including Nazi Minister of War Production Albert Speer) were interrogated. The entire Special Sections Subdivision was discontinued on June 2, 1945, with its functions incorporated into the newly established Field Information Agency, Technical (FIAT), which was a joint US–UK agency designed from its inception to break into national components upon the dissolution of SHAEF. After this happened on June 15, 1945, FIAT became FIAT (US) and FIAT (BR). After much prolonged debate over the desirability of working with the French, FIAT (France) eventually joined these agencies, and each exchanged liaison officers.[20] FIAT (US) was an agency under the purview of the Office of the Military Government, United States (OMGUS), who ran the US Zone of Occupation.

Separate from these was the Joint Intelligence Objectives Agency (JIOA), established by the Joint Chiefs of Staff in 1945 under the Joint Intelligence Committee, making it the bureaucratic brother of TIIC. The JIOA in turn organized Operation Overcast, a top-secret effort to find the scientists and engineers connected with the V-weapon rockets and bring them into American control, among them Wernher von Braun and his team from the Peenemünde rocket testing facility in Germany. When the name "Overcast" was leaked to the public, the project was renamed Operation Paperclip, and its mission expanded to denial of German scientists and technicians to foreign countries as well as acquiring them for the United States. Paperclip operated in cooperation with British colleagues but along military lines largely independent of FIAT or TIIC. Paperclip is surely the best-known of these efforts, and often all efforts to take German scientists get lumped into that name in popular memory, despite it being one small part of this constellation of agencies.

Finally, President Harry Truman issued Executive Order 9568 on June 8, 1945, instructing the Department of Commerce to establish a Publications Board under its Office of Technical Services (OTS), which would be responsible for releasing to industry all scientific and technical information developed by the United States during wartime, pending declassification and national security limitations. Executive Order 9604, issued August 25, 1945, expanded the scope of these orders to include the publication of "enemy" science and technology. Unfortunately, if there are detailed logs of the debates and rationale behind these orders, neither I nor other historians have found them. John Gimbel notes that Fred M. Vinson, director of War Mobilization and Reconversion, was a driving force, especially in lobbying for releasing to American industry not just information produced internally by the US government but also German industrial intelligence.[21] Vinson's proposal to release this information circulated through the State Department, War Production Board, and other agencies as well as the White House, so it presumably received buy-in from a range of executive agencies. The Publications Board, headed by Vinson, collaborated with the Agriculture Department's library, the Library of Congress, and less formally with trade journals to publicize and reproduce reports.

Beyond the aforementioned efforts, there were many smaller efforts operating partly or fully independently with some mandate for investigating German science and technology. The Alsos Mission investigated reports of a German atomic research program, and their capture of the office records (and officers) of the German National Research Council provided the core lists of scientists and technicians from which most other American target lists were built.[22] The Strategic Bombing Survey nominally sought to measure the impact of the bombing campaigns but over time expanded its mission to investigate German industry at large.[23] A Technical Oil Mission, organized by the oil industry in consultation with Harold Ickes, secretary of the interior, investigated German advances in synthetic rubber and other oil-based products. Army Ordnance sent groups to study military advances, as did a US Navy Technical Mission.[24] The original purpose behind FIAT was to coordinate and tame these overlapping efforts. Like many well-meaning attempts to simplify bureaucracy under one umbrella organization, though, it never had the bureaucratic clout to force all of the others to follow its lead, and so it sometimes became just one more competitor adding to the confusion.

As I said, a mess. The organizational charts shown in figures 1.1 and 1.2 are only rough approximations but can perhaps be somewhat useful.

Simplifying the Story

Most of the history of US efforts at technical exploitation can be told by focusing on just a few of these agencies. T-Force units deserve mention as the earliest units to capture German scientists and technical equipment, though their role was short-lived. The most important American institutions were TIIC and FIAT, and they were functionally the same agency, as they shared personnel and had identical missions. The only real difference was that FIAT was based in Germany and TIIC was stateside. Headed by John Green, OTS was the public face of these efforts, advertising FIAT reports to industry and issuing press releases to inform the public. Project Paperclip has attracted by far the most public attention, both in terms of reactions at the time and in popular history about the United States using "Nazi scientists."[25]

In the United States, TIIC (and later Green's OTS) communicated with industrial leaders and trade associations to identify targets worth investigating in Germany, and then to recruit technical personnel from these firms to

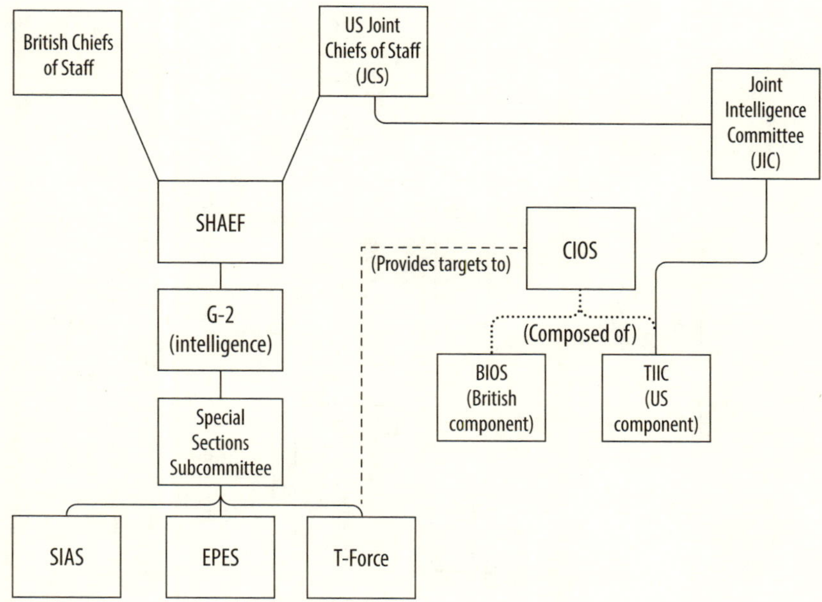

Figure 1.1. US agencies studying/taking German technology, SHAEF era (before July 1945).

be investigators on two-to-three-month tours. Upon arrival in Germany (or more often, in London on the way to Germany), these investigators received basic instructions, a faux uniform, a nominal military rank equivalent to colonel, and introductions to T-Force units who would handle their transportation and housing. Teams of investigators with related interests would travel to preapproved facilities, where they had authority to question technical personnel and managers, copy (but not remove) any paperwork, and tag machinery for reparations seizures. Upon returning to the United States, investigators would write up reports about their findings, which the Publications Board would publish and publicize, allowing their acquired knowledge to benefit all US industries and indeed companies around the world (the Commerce Department sold reports to all interested, and many foreign firms and countries purchased copies). Investigators wishing to travel to the US or French occupation zones would apply for passes through the FIAT liaison officers, and then the French or British authorities would care for them during their visits.

This was all theoretical, of course, and in reality things did not work quite so smoothly. Some individuals or teams of investigators traveled to unapproved facilities, exploiting these "targets of opportunity" despite FIAT's ob-

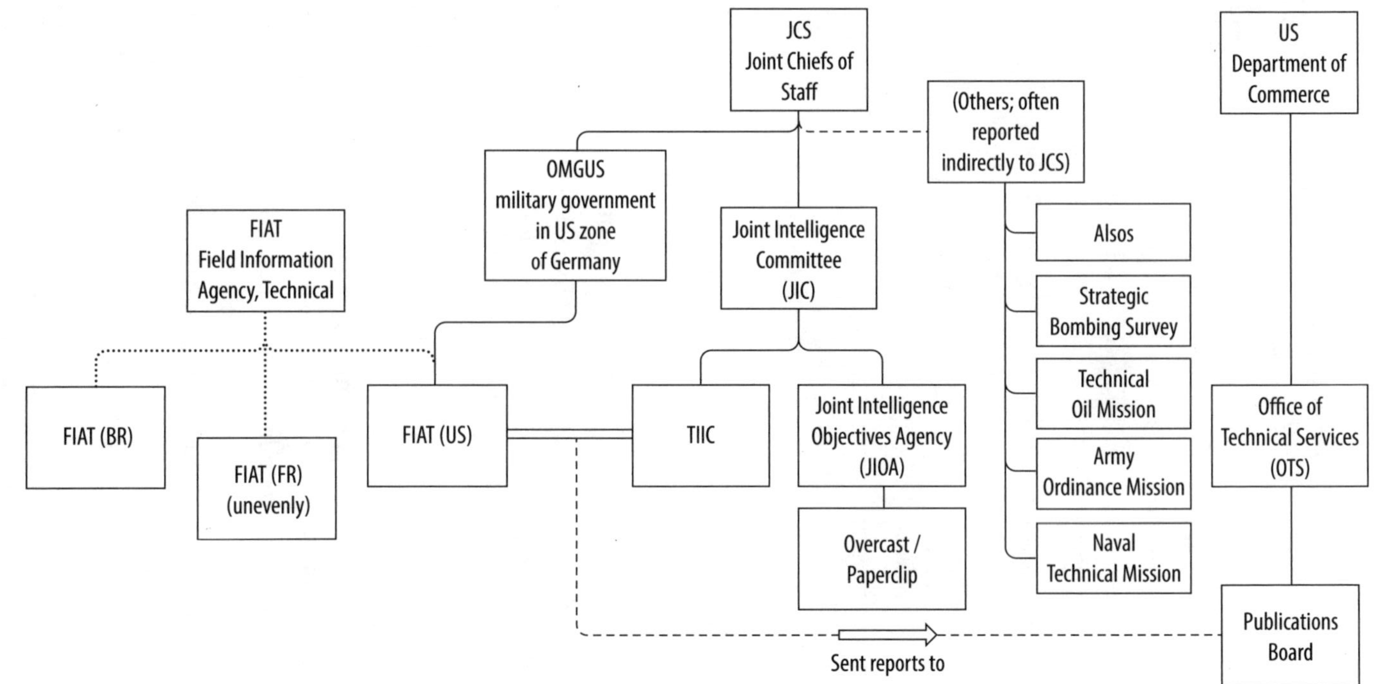

Figure 1.2. US agencies studying/taking German technology, post-SHAEF era (after July 1945).

jections.[26] Many investigators never bothered to write their final reports or wrote terse and unhelpful summaries that provided little detail.[27] Transportation and communication were major bottlenecks, as FIAT lacked the authority to command the military government officials to provide resources. Though initially enthusiastic about the founding of FIAT to coordinate efforts in Germany, TIIC's chairman eventually became convinced that its lack of clout in the OMGUS bureaucracy actually made their job much more difficult.[28] Only frequent appeals to President Truman's support—and through him, that of top generals—eventually secured FIAT access to offices and trucks.[29] In most reports, the Germans who were interrogated were helpful and open. In some cases, however, they were not so helpful. Interrogators had no authority to employ physical violence, but scientists who refused to cooperate often ended up in jail, or at least under close surveillance. The three Western occupying powers generally cooperated with one another's investigations, but there are scattered reports of French teams covertly hiring German scientists being kept in American or British camps for future employment or technicians being put into "protective custody" (i.e., jail) for threatening to flee to other zones with their industrial knowledge.[30]

Despite these day-to-day difficulties, hundreds of investigators from the United States joined thousands from the United Kingdom in touring Germany from the end of the war through 1949, and scientist "denial" programs continued to operate long afterward to various degrees.[31] Hundreds of thousands of documents went through data processing centers, only a fraction of which were judged valuable enough to be worth translating and releasing but which resulted in thousands of final CIOS, FIAT, and BIOS reports.[32] These, in turn, sold quickly. By January 1950, BIOS (the British equivalent of FIAT, though a FIAT [BR] did exist for liaison purposes) had dispatched more than 46,000 copies of summaries and abstracts to industry.[33] American efforts, as we will see in chapter 6, were even more ambitious in scale, microfilming far more documents than they would ever even attempt to translate, much less distribute.

Exploitation Programs on the Ground

Reviewing memoirs and letters by these investigators lets us cut through some of this high-level bureaucracy to see what FIAT looked like on the ground. Two investigators, Gunther Stent and Nelson Leonard, make a striking pair for these purposes. Both were research chemists at the University of Illinois at Urbana-Champaign (Stent a doctoral candidate, Leonard a postdoctoral researcher) during the war. Stent worked on synthetic rubber, Leonard on synthesizing antimalarial drugs for use in the Pacific Theater. Eventually, both would go on to lead tremendously successful scientific careers. Stent went on to work with Max Delbruck at the California Institute of Technology as part of a "phage group" that made fundamental contributions to molecular biology, and then became a professor at the University of California, Berkeley. Leonard became a professor at the University of Illinois, where he helped found the field of bioorganic chemistry and was elected to the National Academy of Sciences.

Both took leave from Illinois between 1945 and 1946 to serve as scientific investigators for FIAT in Germany. Despite the similarities in their positions, their recollections of FIAT's value are tremendously different. Using two investigators' memories out of hundreds means their stories are not necessarily representative, of course. Still, as we will see, there is a substantial split in the historical record between those who raved about FIAT's value and those who saw it as a waste of time. These two investigators, then, can give us an on-the-ground view of some of FIAT's work while introducing some of the challenges of interpreting its legacies.

Perhaps not coincidentally, Leonard's mentor at the University of Illinois was Roger Adams, who served as the chief consultant to OMGUS on how to control and revive German science, and to whom we will return later in this book. Leonard served in FIAT from September 1945 to February 1946. His recollections paint an exciting and productive adventure. After arriving in Germany and settling into the document library and microfilming facility at Griesheim, Leonard set to work reviewing technical files. The nature of the work fit his concept of how to study science—rather than spending time interrogating German technicians, "it was my contention that the material of real transfer value lay in the research reports and process directions. These we gathered in from all sources, starting with the separate plants belonging to the I.G. Farbenindustrie, evaluated them, indexed them, and had them microfilmed for transfer to the United States. . . . The team, which consisted of 28 personnel during the initiating phases while I was in Germany, worked very efficiently as long as the scientists and translators did not stray from their labors."[34]

Leonard quickly discovered the value of efficient archivists. A British counterpart recommended that FIAT "hire a German librarian if I wanted to have my (research) files 100% complete. I convinced the commanding officer in Hoechst to do just that on the basis of Commander Child's experience, and it had a remarkable effect on the operation of our document center. Missing years of research reports appeared as if by magic. I never asked about their sources when they appeared suddenly, usually following a weekend."[35] After his tour, Leonard returned to Washington, DC, sat through a debrief by JIOA under the Department of Commerce, and flew back to Illinois just in time for the spring semester to begin.

Stent's experiences were much less flattering to FIAT and much more in line with the challenges highlighted later in this chapter. Stent was a Jewish refugee who had fled the Nazi regime in his youth, eventually becoming a US citizen. He heard about FIAT while reading *Chemical and Engineering News* and saw in it a chance to visit his childhood home (and, as he writes in his memoirs, to gloat a bit over the suffering of the Nazis who had persecuted his family and his community).[36]

Stent arrived in Germany in late 1946. During his orientation, Stent's immediate superior, "Freddy K.," a Czech-Jewish engineering student and US Army sergeant, described FIAT's history as follows:

> We got started right after VE Day. During the first year of our operations, field investigators were mainly volunteer hot-shot technical experts, dollar-a-year men on paid leave from American industry. They knew exactly what they were looking for and usually found it. For instance, they turned up new methods for making synthetic rubies, or synthetic gasoline, or synthetic rubber. Once one of those industrial hot-shot investigators had found the novel method he was looking for, he took it back Stateside, to have it put into production at savings of millions of bucks in development costs.
>
> These company types passed on the nitty-gritty working details of the German technical breakthroughs only to their own firms. They usually hid them from their stateside competition. When word got out about this, Washington decided that, from now on, FIAT investigators were going to be paid employees of the Department of Commerce. So this year, they hired a new crew of people's-own investigators, like you. You're supposed to be working for the nation, and not for the moneybags at DuPont, Firestone, and Upjohn. There's another difference too. You guys won't be sent out to track down the specifics of particular pieces of novel technology. You're supposed to liberate all the hot technical stuff you can find.

Stent's first question might well mirror ours today: "How am I supposed to do this? Go from one place to another and simply say to the Kraut in charge, 'Tell me all about the secret technical information you have, Mister!' What if he lies, like 'Frightfully sorry, Sir. I haven't got any secret technical information'? Am I supposed to extract the truth by torture?" To Sergeant K's explicit regret, torture was not allowed, so FIAT investigators were forced to simply "carry out [the] vacuum-cleanerlike intelligence mission only by looking at technical documents."[37]

Stent's team of investigators assigned to IG Farben (one of the largest, oldest, and most powerful chemical combines in the world) represented the diversity of these "peoples' own" investigators: "a young Austrian-Jewish refugee, who was, like me, a stateside physical chemistry graduate student; a middle-aged, scientifically and linguistically unqualified Good-Time Charlie from North Dakota; and an elderly, taciturn Austro-Jewish couple from New York with vague scientific credentials."[38]

In Stent's memoirs, the IG Farben Leverkusen plant director-general was more than accommodating in providing access, but another problem immediately presented itself:

I asked my colleagues, "How are we going to tell which documents describe hot technical information? And which are merely old hat, already known all over the world? Take me, for instance, a student of—let's even exaggerate and say, an expert in—the physical chemistry of large molecules. How am I going to decide whether a procedure for making some drug of which I've never even heard, is hot stuff? We'd have to be some kind of universal geniuses to do a real screening job on all the paperwork piled up here!"

The other physical chemistry graduate student chimed in: "Yeah. And even if we were universal geniuses, what with that huge pile of documents we are supposed to screen here, it'd take us forever. We'd all be still sitting here in this comfy office at the turn of the twenty-first century, still slowly turning the pages of loose-leaf binders. Like Emperor Friedrich Barbarossa sitting in his Kyrfhaeuser Mountain cave over the centuries, waiting to save the Holy Roman Empire, while his red beard grows through his table!" Good-Time Charlie also had a good point: "How are we going to keep our Kraut camera crews busy? It takes a lot longer to read a document than to microfilm it!" Thus, before even getting started on any screening, we cottoned on to the futility of our mission, and, indeed, to the hare-brained nature of the whole FIAT document-screening program.

The team's strategy was to sidestep the issue: they would simply mark every shelf with documents produced in the past twenty years as needing to be copied. "This brilliant rationalization of our screening procedure left us with a lot of spare time for extracurricular activities and resulted in the exposure of enough 35mm film to stretch from Leverkusen to Washington."[39] As we will see, these were far from the only investigators to ask these questions, and solutions were often as half-hearted. In the American case, though, this was never as central of a concern as it became for the British for a simple reason: American firms who could afford to send investigators to Germany often found they were not particularly interested in replicating what they found back home.

Washington Politics and the Distribution of German Science

In Washington, political wrangling put its own imprint on how science and technology found in Germany (and produced by the US government) was shared with US industry. The OTS bore the brunt of these political struggles. Personal rivalries, anti-communist paranoia, small-government ideology versus New Deal progressivism, and basic Democratic versus Republican angling for party control all made for tumultuous early years for the OTS. In turn, it was generously funded, then strung along, gutted, rebuilt, and eventually transformed into an embodiment of government taking a new, powerful, leading role in scientific information systems.[40]

As mentioned previously, Truman's Executive Order 9568 in June 1945 sought to release to the public as much of the scientific research that the government had funded during the war effort as possible, within the constraints of national security. Executive Order 9604, issued a month later, expanded this mandate to include information on German science and technology. This duty fell to the secretary of commerce, Henry A. Wallace, formerly one of FDR's most ardent New Dealers as secretary of agriculture—and a man with many political enemies, both personally and as a New Dealer. Wallace created the Office of Declassification and Technical Service, which would later become the OTS. Wallace was among a set of progressive policymakers who saw the OTS as potentially filling a long-needed role: a government hand in connecting businesses with science and technological improvements. If successful here, he hoped, it might become a permanent role for the government, aiding small businesses first and foremost. For many in Washington, the OTS's successes and failures were a proxy for Wallace's, whatever the actual merits or flaws of the OTS's mission.

John Green, formerly the chief engineer of the National Inventors Council, took charge of the OTS and coordinated with the Departments of War and the Navy on which technical reports were suitable for release. As I discuss in detail in chapter 6, the OTS was at the heart of a major, lasting shift toward more government involvement in distributing scientific research. The OTS's own vision of its mission centered on helping small businesses, though large firms were also obviously intended beneficiaries. The logic was similar to what British planners were thinking at the same time: large firms (e.g., General Electric or DuPont) could already afford research and development, and so it was the companies without those capacities who stood to gain the most from investigations into German industry. More broadly, Wallace and Green saw the OTS building outward, as a New Deal–inspired government would organize and aid the flow of scientific and technical communication in the postwar world, ensuring American industrial competitiveness.

Despite the OTS's early popularity and good press, including very positive coverage in the *Wall Street Journal*, *Newsweek*, *Science*, *Business Week*, and *Harper's*, it quickly ran into political trouble, due more to personal and party politics than to objections about its mission or methods. Wallace's vision of the OTS as a step toward a more active government role in science communication drew opposition from Republicans who sought to scale down the government bureaucracy drastically now that the war was winding down. They feared that the dramatic increase in government powers necessitated by the war would become the new normal. Meanwhile, Wallace himself had enemies even within his own party. In the 1940s and 1950s, the Democratic Party was a coalition that included both conservative Southern Democrats ("Dixiecrats") and idealistic, progressive New Dealers. The Dixiecrats feared that Wallace would be a threat to the reelection of southerner Harry Truman and hoped to undercut his success. Both sets of opponents benefited from Wallace's stance of collaborating with the Soviet Union on nuclear technology, which made him a prime target for anti-communist hysteria in the late 1940s and early 1950s. The OTS became a pawn in this larger political game.

In both 1945 and 1946, Senator J. William Fulbright presented bills that would have established the OTS permanently. After very positive hearings full of encouraging testimony, Senate leadership nonetheless declined to advance the bill to the floor for a vote. Indeed, not only did congressional Republicans sabotage efforts to institutionalize the OTS as a permanent entity, but they actively sought to defund it entirely. In part, this played out in a broader effort to defund institutions of all kinds built during the New Deal and then during the war, and the Department of Commerce as a whole was a target. The OTS, as an explicitly war-related entity (to share the fruits of science developed by the US government and in Germany during the war), drew particular ire. Several times, including in 1947, the House removed any funding for the OTS from their appropriation bills before conceding some limited funds (790,000 USD in 1947) in a conference committee compromise with the Senate. In 1948, the OTS budget was cut to 200,000 USD, with explicit suggestion that it might receive nothing the following year.

In 1949, Democrats swept into power in Congress again, and the prospects of the OTS receiving permanent status rose with them. After a year of hearings and negotiations, Congress passed and President Truman signed Public Law 81-776 on September 9, 1950, to "make the results of technological research and development more readily available to industry and business, and to the general public, by clarifying and defining the functions and responsibilities of the Department of Commerce as a central clearinghouse for technical information which is useful to American industry and business."[41]

In chapter 6, I go into more detail regarding the importance of the OTS and its successors in changing the landscape of international scientific communication, which is one of the most important legacies of the investigations into German science and technology. On the immediate issue of American exploitation programs benefiting US industry, though, this revival was somewhat too late. By February 1947, John Green was issuing a "Last Call for Germany" in the *Federal Science Progress* (the primary OTS publication) as well as in any trade journals that would republish it, advertising a last chance to send investigators overseas.[42] The budget cuts of 1946–1948 removed any possibility of thoroughly translating and indexing the vast quantity of primary documents and summary reports arriving back from Germany. The vast majority ended up in long-term archival storage, where they were neither translated nor possibly even looked at again, with rare exceptions for curious historians.

(Relatively) Concrete Gains

Before diving into what did *not* work regarding American exploitation of German science, it is worth looking at the real, important successes. In at least some clear cases, the technologies and personnel that the FIAT-related agencies brought to America had a well-defined economic, political, or social impact. I argue that John Gimbel likely overstates the case when he claims that "the $10 billion figure bandied about by the Russians and their friends and dismissed by State Department functionaries as 'fantastic' is probably not far from the mark," but neither should we ignore the financial and national security significance of these programs.[43]

If the general public knows one name associated with German ("Nazi") scientists benefiting the United States, it is undoubtedly Wernher von Braun. He and his team of rocket scientists designed Germany's V-2 missiles at the Peenemünde facility and were some of the most prized targets of Project Paperclip. After arriving in America and finding work with the army and then NASA, von Braun became famous in the 1950s as a proponent of manned spaceflight. He and his story became even better known after his team aided NASA in designing the Saturn V rocket used in the moon landings. Since then, he has inspired film characters (e.g., the chief scientist in *The Right Stuff* and the title character in *Dr. Strangelove*), documentaries, and even music parody by Tom Lehrer ("'Once the rockets are up, who cares where they come down? / That's not my department,' says Wernher von Braun").[44]

Von Braun did not come alone. When he first approached American forces to negotiate a surrender, he was clear about wanting to bring the entire group with him, even exaggerating the scientific accomplishments of some junior members. About two dozen German rocket scientists ultimately settled in and around Huntsville, Alabama, where they adapted to the local politics and culture. In part due to von Braun's celebrity, Paperclip itself has entered popular memory, inspiring history books and History Channel conspiracy theory documentaries.[45] Some of these popular works are quality scholarship, tightening the case that American officials knowingly employed or protected German scientists guilty of war crimes, including bringing zealous Nazi doctors who had performed unethical experimentation, a clear war crime, to America.[46] Other portrayals are more focused on scandal than nuance and equate "German scientists" and "Nazi war criminal scientists," when the role of science and scientists in Nazi rule was very complex.[47] Quite often, the writing confuses Paperclip with the broader set of programs discussed here.

What exactly America *gained* from Paperclip beyond the von Braun team is rarely clear or concrete, but there are a few indications. Most of the scientists brought to America were housed in air force facilities at Wright Field in Ohio. While there, from February 1946 onward, the air force allowed businesses to interview these specialists and sometimes even "loan" them out for months at a time.[48] Lockheed, Douglas Aircraft, Westinghouse, General Mills, Boeing, and Bulova Watch Company are among the companies who expressed interest, and General Electric at one point indicated that consultations from one specialist might have saved the company about 1 million USD. Some of this interest seems to have been based on Germany's reputation, such as Bulova Watch Company's discovery that the specialists "did not in fact have special knowledge desired," but "Mr. Bulova himself had taken an active part in getting these men brought to this country by the Army," so they hired the Germans anyway.[49] Accounts vary from Paperclip scientists being useful in other particular industries to others who the army struggled to place in industry at all. We should not extrapolate too far from von Braun's importance to assume that German aeronautic technology was so uniformly vital to American industry.

Still, in terms of the space program itself, von Braun's leadership in developing the Saturn V rocket seems to have been crucial. In the words of historian and von Braun biographer Michael Neufeld, "the efforts of the growing corps of scientists, engineers, and technicians . . . would have been wasted but for von Braun's superb technical leadership."[50] Whether another—American— leader would have stepped up and achieved the same in a reasonable time frame (even if a bit slower) is hard to say, but von Braun and his team seem to have contributed substantially.

The V-1 and V-2 self-propelled bombs (and the space race they foreshadowed) were not the only German innovations in aeronautics coveted by the US military. The air force (or, more properly, the US Army Air Force, as it was not a separate branch of the armed forces until the National Security Act implemented in September 1947) was also extremely interested in German advances in jet fighters.[51] The possibilities of jet engines were clear even before the war, but producing them in practice, in quantities and with reliability to be relevant in war conditions, seemed beyond US capabilities. As a result, they had decided it was better to build from a more developed British design, similar to how they had borrowed and built on British radar and penicillin technologies. They never reached meaningful production before the end of the war. In an increasingly desperate and resource-starved Nazi military production system, meanwhile, the relative safety and reliability of piston engines were less highly valued than jet engines' potential for cheap, fast production and their use of diesel fuel. As with the V-2, Nazi jet engine design, production, and testing depended extensively on brutal slave labor.

Aircraft development and testing gear is one of the few areas where the United States was especially eager to receive physical reparations, as German wind tunnels were far superior and more numerous than what American industry or military had to offer. During the Nazi era, Germany built sixty-two wind tunnels, compared to three in the United States.[52] Unsurprisingly, the air force and US aircraft firms such as Boeing were eager to import both research scientists involved in designing jet engines and ground-level technicians experienced in assembling and running these wind tunnels. Operations

Overcast and Paperclip mostly focused on these aerospace personnel. In another symbol of the prewar ties between American and German science, General Henry H. ("Hap") Arnold put the US Army Air Force Scientific Advisory Board under the command of Theodore von Kármán, who had studied, taught, and researched in Gottingen and Aachen in the 1920s. The air force, guided by this Scientific Advisory Board, pushed for more and more aeronautical personnel—ultimately, hundreds—to be brought into the United States.

Assessing exact contributions is difficult, but on the whole, American investigations into German aircraft technology seem to have had real returns, albeit ones we need to keep in context. Some testimonies, such as that of North American Aviation in 1947, indicate huge research savings.[53] However, many of the German scientists brought to Wright Field, the air force staging facility for this process, ended up being of minimal interest to US industry, and their value (if any) came from denying their expertise to other nations for a time. Keeping in mind that British innovations during the war were also extremely important in the longer-term process of developing a functional, reliable, mass-producible set of aircraft technologies (including supply lines and infrastructure), the air force and US military seem to have gained significantly from extended contact with German research and expertise in aircraft design and production.

The German chemical industry was world-renowned by the start of the Second World War, so it is unsurprising that there is evidence of important gains in sections of that industry. A magnesium expert at Dow Chemical commented that "in the magnesium industry the Germans were well advanced and entirely competent and in possession of information which can be profitably utilized in this country."[54] John Green, testifying as head of the OTS, used excerpts from oil industry company leaders to argue that FIAT deserved more funding. Chemical firms such as Standard Oil sometimes pushed back against his overly selective use of praise, noting that many of the basic technologies they found in Germany were already in use in the United States even before the war. Still, they admitted, they had found some very useful innovations, including some important ones.[55]

Synthetic gasoline and rubber were technologies that were never economical in the prewar US economy. It was cheaper to use American oil, or to import oil, and the United States had at least semicolonial relationships with Southeast Asian countries and colonies that supplied abundant, cheap rubber. Once war began and Japan cut off these East Asian suppliers and German submarines threatened oil shipments across the Atlantic, American industry experienced a supply shock. Germany, meanwhile, had never had as much access to these foreign suppliers. With a strong push from a Nazi government ideologically insistent on being self-sufficient, German science and industry had led the world in transforming coal and other local feed stocks into rubber and gasoline. The Fischer–Tropsch process, in particular, was a breakthrough in producing synthetic gasoline and had only started to spread around the world before wartime embargoes cut off most access.

American firms had heard about German advances prior to the war, of course, through trade journals and personal contacts. As they struggled with shortages, American firms (and the military for whom they were contracting) placed a new premium on self-sufficiency and were eager to learn more about the rumored German advances. The Petroleum Industry Council, an industry group organized by the US Department of the Interior to coordinate war production of oil, was able to lobby successfully for a special "Technical Oil Mission" to follow just behind the front lines during the invasion of Germany.[56] The industry recommended twenty-six of its top synthetic fuel experts to serve as investigators in August 1944. It took another six months to get the mission organized and off to Europe, but they reached the ground by February 1945.

The participants of the Technical Oil Mission copied tremendous amounts of information and had the opportunity to interview German specialists from leading firms. However, there is an enormous difference between copying files and "taking" a technology. As the war ended, Middle Eastern petroleum and Asian rubber once again flowed into US markets. As prices dropped, so, too, did industrial interest in synthetic oil and rubber. Most of these findings remained untranslated, stored in boxes in bulk, until the oil crisis of the 1970s spiked interest in synthetic gasoline once more. At that point, the German Document Retrieval Project at Texas A&M University sprang up to attempt to gather, translate, and disperse these thousands of boxes of files from dead storage around the United States, and thereby to resurrect this half-forgotten technology.[57] The end of the oil crisis again dampened enthusiasm for synthetic oil.

What, then, did America gain from the Technical Oil Mission? There was some assurance that if the Cold War had somehow entered a nonnuclear total war, America would have theoretically had the ability to produce gasoline in bulk from coal stores. That had real national security value. Germany in this case lost very little, aside from skilled chemists being discovered and offered better jobs in the United States. Even then, given the level of unemployment and desperation for resources in occupied Germany, having German citizens earning relatively very high wages abroad that they could send home had value on its own, if offset by the brain drain potentially harming long-term recovery.

A few individual innovations have been documented as having entered the US economy through FIAT. One example is magnetic tape for audio recording.[58] While magnetic recording was invented in the 1880s, it caught on much faster in Europe, since American radio used live programming to a greater extent. In the 1930s, the German firm Allgemeine Elektricitäts-Gesellschaft (AEG) produced the Magnetophone, a device for high-quality broadcast audio recording. Magnetophones became standard across Europe after the Nazi broadcasting authority adopted the model, and FIAT brought several prototypes and reports about the technology back to America. Several US firms who ordered the FIAT report on the Magnetophone introduced their own models soon afterward, including Rangertone (founded by the author of the FIAT report), Ampex Corporation, and Orradio Industries. Of the four companies who produced American versions, only 3M did not have direct connections to FIAT.

This still leaves unanswered the important question about what FIAT accomplished in regard to these inventions (of which the Magnetophone is just one example among many) *that would not have happened otherwise.* That is, without FIAT, would magnetic tape recording have become important in the American marketplace? Some firms very clearly benefited here, and that is worth noting, but how far can we claim that the *economy* benefited from this technology being more easily available to American firms? History cannot, of course, definitively answer "what if" hypotheticals, but we can make reasonable inferences and ask readers to come to their own conclusions. The existence of the 3M model (which does not seem to have any clear connection to the investigations in Germany) shows that well-off US firms could and did develop similar products. Magnetic tape recording was superior in some technical senses but required changes in the overall US radio broadcasting infrastructure before it really made economic sense. The technology has a chicken-and-egg / critical mass problem familiar to many tech start-ups today.

American industry gained *information* about Magnetophones from these investigations (again, used as one example of individual technologies brought back), and information is valuable. Still, we should both appreciate what the investigations *did* accomplish and keep in perspective the extent to which they truly enabled something that would not have happened otherwise. Without these investigations, it seems likely that fewer firms would have produced Magnetophone technology, and that might have meant higher prices and a slower (or nonexistent) adoption of the technology overall. It is these kinds of marginal effects that we would need to stack up against the price of administering and running FIAT if we wanted a dollar value accounting of the benefit for the US economy. Similarly, what was lost for Germany? Possibly AEG lost some of a potential US market. Conversely, the US radio broadcasting industry did move toward AEG's standard in part because of these investigations and the ensuing American competition, in effect *developing* a new, potential market and making AEG's products a new kind of standard throughout the United States and Europe. Here, too, the marginal effects are real and important. They are not, however, the zero-sum game of physical war booty such as paintings or silverware, where benefiting one party means depriving another equally. Keeping this balance in mind is crucial in any discussion of technology transfer and "intellectual property"—itself a relatively recently accepted legal term that we should be careful not to take too literally.

The General Case: Disappointment in German Technology

The FIAT investigations in Germany did not just target a few specific technologies with major wartime innovations, such as chemicals or rockets. Teams of FIAT investigators wrote reports about toy making, wood harvesting, watch production, machine tools, cottonized flax, gear manufacturing, textiles, processing of fats and oils, centrifugal casting of metals, mining, surgical equipment—a huge variety of technologies.[59] Most investigators, especially in the early stages, were on loan from private industry, paid by their employers rather than by the government. That means not only must policymakers have thought this was worth it, but businessmen in each of these many industries were willing to put their money and strained resources behind it.

So was German technology actually better? It is a fundamental question for judging what America got out of these programs, but it is a tricky one. You could argue that the existence of these programs is itself evidence that they must have been worthwhile—why else would people get so excited over the chance to learn from German industry? Historian John Gimbel stops short of making a claim that German technology was uniformly superior to what American industry already utilized but "accept[s] what Vannevar Bush and others more qualified . . . have had to say . . . [that] modern industries developed variously and unevenly, and in this particular case Germany was ahead in certain areas [of industry] while the Americans led in others."[60]

This is not a story, ultimately, of the United States getting a free lunch at Germany's expense (or at least an enormously valuable one). Instead, it is a story of American firms having an unprecedented ability to study a long-standing rival, realizing that they had relatively little to learn and moving forward as self-aware leaders in industrial research. This, in turn, led to a reorientation of the West's scientific and technical communities. In most industrial fields, for most investigators, the biggest surprise to be found in Germany was not its host of scientific secrets. The biggest shock was one of disappointment, the discovery that American and British technology was, by and large, more advanced and more suited to their countries than anything to be found in the land of (as British policymakers put it) "ingenious barbarians."[61]

American businesses had been eager to join in and volunteer technical investigators, and as the programs approached their ends, TIIC/FIAT reached out to them for feedback. Had they benefited? How so? Many never responded, perhaps unsurprisingly. Those who did frequently gave a strangely contradictory assessment: the program was great overall, sure to be of tremendous value, but in their own field there had not been much to learn. "I found nothing, in its entirety, acceptable to us and our industry," wrote one chain manufacturing company investigator, yet later he added: "Retrospection has provided the conclusion that the trip was of considerable value, especially to my company, and that the value was in the mechanical designs and details in tooling, no single one of which might be considered of consequence."[62] "Germany's advances in her war-time automotive industry do not measure up to those of America," began another investigator, and they "can never hope to surpass America." Yet he concludes, "What FIAT is doing—what FIAT is finding—will be of inestimable value to American engineers and industrialists. American industry is surely not taking full advantage of this government service. . . . That is one thing that has surprised me greatly."[63] An investigator in high-pressure hydraulics who had expected German technology to be "far advanced over America" instead found that German industry employed equipment that had "a much lower safety factor than does American industry"— yet he rated his trip "well worth while."[64]

We might expect the records kept by FIAT and others (now stored in the US National Archives) to reflect a somewhat rosy picture of the agencies' value. Those who gained little from the investigations seem less likely to have responded to follow-up requests, and the administrators in these agencies had an incentive to seek out positive testimony, especially as fodder for the ongoing congressional fights over defunding the OTS. Looking beyond the records kept by the exploitation agencies unveils sharper skepticism. An article in the *Christian Science Monitor* in March 1946 describes the disappointment felt by the membership of the Society of Automotive Engineers upon hearing from its investigators:

> [An investigator] had heard so much "propaganda" from Germany that he had almost been convinced that they had something quite superior to offer, but that his visit had made him disappointed in the accomplishments of the Germans. . . . All evidence indicates . . . that German vehicles were generally inferior to our own in point of dependability and relative freedom from troubles. All in all, meetings indicated that it was time for American engineers to drop all feeling of inferiority left over from the days when it was conceded generally that as craftsmen and engineers Germans had no equals.[65]

Trade journals for key industries targeted by FIAT contain many statements skeptical of value. In *Automotive and Aviation Industries*, a March 15, 1945, article surveying production techniques in the German aircraft industry commented that "the quality of sheet metal work . . . is considerably below the standards of United States manufacturers," "a very evident lack of knowledge of the use of chip breakers in tool grinding was noted everywhere," and "machine tools are very similar to those of the United States," even to the extent of German machines being copies of the American-made Brown and Sharpe automatic lathes.[66] A report written in May 1946 honoring the FIAT investigators emphasized their bravery and the importance of their mission yet mentioned the "general impression . . . that enemy machine tools were not up to our standards and held little of value to machine designers."[67] The article "Nazi Reports Disappointing" in *American Machinist* critiques FIAT reports as "interesting, but not one in ten contains data of use to American industry."[68] And an issue later: "Generally speaking, there appeared to be no outstanding developments [in German machine tools] apart from normal improvements."[69]

In industrial health, medical devices, and scientific instruments, the results were similar. In September 1945, the journal *Industrial Medicine* included a letter to the editor from Colonel Edward D. Churchill, writing about his tour of six German military hospitals: "There was considerable expectation that the German doctors, with the German medicine's world-wide pre-Hitler fame and the well-known German thoroughness and energy, would have some pretty phenomenal achievements of their own to report from their war hospitals," but he found German methods "about 20 years behind the American procedure."[70] One investigator for the American Instrument Company gave talks on his experiences to the Scientific Apparatus Makers of America and found great interest there—but "the 'human interest' phase of the trip was of much more interest than the technical phase."[71]

The chemical industry is one in which German firms, including the famous IG Farben cartel, certainly lead the world in at least several important technologies, and indeed German chemical industry investigations received heavy coverage in American trade journals. The editor of *Chemical and Metallurgical Engineering*, Sidney Kirkpatrick, was an eager booster for FIAT, writing several editorials encouraging industrial cooperation with FIAT and use of its reports. Both this journal and *Chemical and Engineering News* reported extensively on German processes in multipage detail, including a bibliography of newly available FIAT/BIOS reports in each issue. They referred to these reports as being "of tremendous interest to chemical engineers in this country," "anxiously awaited by the US chlorine industry," and "a remarkable development of an acetylene industry," and FIAT as "one of the most beneficial programs for American science and industry," with some of these articles written by representatives from companies such as Dow Chemical and Tennessee Eastman Corporation.[72]

Even in chemicals, however, there are signs of sincere disappointment in German science and technology. As just one example, an editorial in *Chemical and Metallurgical Engineering* from July 1945, sardonically titled "Ueber Alles?," begins: "Once more myth of Germany's well-advertised superiority in chemical matters has been exploded by the reports of the mission of American technologists who inspected the Buna plants and laboratories prior to V-E day. . . . There is no indication that German synthetic rubber techniques will be of material help to the American rubber industry because our present processes are superior in so many respects."[73]

A September 1946 editorial in *Technical Services*, a newsletter of the Department of Commerce's Publications Board, began to qualify initial estimations that German technology was broadly superior: "Because OTS collects and publicizes so many German technical developments, we are often asked if we believe in the 'superiority' of German science. Returning investigators agree that German science and technology were, in most respects, far behind ours. In seeking to meet the demands of the military in metallurgy and aeronautics, and in attempting to find substitutes for petroleum and other critically short materials, German scientists produced inventions, ideas, processes, and formulas which were unique, outstanding, and valuable. American industry can and will benefit from examining these developments and adopting some of them." If the investigations were simply seizing an easy opportunity, you would expect the same to be true for Japan. Like Germany, Japan faced long-term occupation, and investigations could proceed without worrying about British, French, and Russian occupation zones. However, the same policymakers decided from the start that "any large-scale exploitation of Japanese science and industry would not be a justifiable expenditure of government funds."[74] The Technical Industrial Intelligence Division, the same basic apparatus as TIIC but renamed after its move to the Commerce Department, sent out questionnaires to American industry asking whether they believed investigations of Japanese industry were worthwhile, and if so, what should be researched. Most responses were noncommittal but supportive. Boeing, Spencer Thermostat, and many others expressed belief that "investigation in Japan might be very beneficial," though few were terribly eager. American Airlines, like many others, was "not too keen at the present moment."[75] Ultimately, a few "scouting" trips were organized, particularly in fishery and boat-making industries, but nothing anywhere approaching the scale of the FIAT programs.[76]

This disappointment in German science and technology was not solely an American phenomenon. In the United Kingdom, the archives of the agencies involved in scientific exploitation, trade journals, and press share the initial enthusiasm of Americans, followed by some expressions of sharp disappointment mixed in with the reports of great value acquired. The document processing centers for items copied by investigative teams reported that "the number of documents in any batch which are of real value to industry is very small—possibly not higher than 5%."[77] In a House of Lords debate on a bill to protect users of BIOS reports from copyright and other intellectual property lawsuits, Lord Edward Jessel fought the measure on principle (he saw personal property, including intellectual property, as off-limits even during total war), noting: "I think it is now generally agreed that the results were disappointing, and that although the reports of the teams may have infringed copyrights, they added little to our industrial knowledge."[78] The Association of British Chemical Manufacturers wrote to the head of BIOS in October 1946 that "the number of really 'novel' processes is comparatively few."[79] A former head of CIOS and BIOS praised the German tradition of ingenuity in producing synthetics and ersatz products but noted that this meant that many of their developments "were ones which we would not necessarily want to follow."[80] It should be said, however, that these disparagements of German science and technology are much rarer in British sources than in American ones, and British expressions of frustration with BIOS originate much more frequently with its methodology and efficiency (as we will see in chapter 2) than with the lack of potential for German technology to be useful.[81]

In July 1947, FIAT sent a final report of its activities and accomplishments to the OMGUS chief of staff. Perhaps reflecting the priorities of the military government to which this report was sent, the first paragraph highlights a very different take on FIAT's value—rather than being "all take and no give, . . . it is sincerely felt that some phases of the FIAT program . . . have been and will be of tremendous importance to the revival of German science and perhaps, to a lesser degree, to economic recovery." German industry, it argued, might have acquired "some small gain . . . from visits by American investigators since, through this means, German industrialists have gathered some insight into parallel activities in the United States and thus have gained a better idea of what the German concern can best or most economically do on a world market." The value to the United States, in contrast, "may be a moot question." If nothing else, it aided the War and Navy Departments in securing war booty such as "rockets, war chemicals, aircraft . . . and wind tunnels." Counted in terms of the "reasonable percentage of data resulting from expenditures by Germans in research," its value in reparations to US industry "is measured in the billions of dollars. It would perhaps be not far wrong to consider that the US Government and industry will financially receive 1000 times more value than it expended in the project."[82] However, this assessment included physical items (e.g., wind tunnels, prototypes, and scientific tools such as precision optical equipment). As this estimation came from Ralph Osborne, the former head of FIAT, there was reason to play up the agency's value. Even here, though, Osborne admitted that "the general impression was that American industry with its massive production lines and high degree of mechanization was far ahead."

We should, of course, take seriously the possibility that the sources disparaging their findings in Germany had ulterior motives. There is real evidence, for instance, that a mix of nationalism, self-promotion, and not-invented-here syndrome in American industry might have incentivized businesses to downplay the value of German developments, even as TIIC and similar agencies sought to justify their existences through inflating the investigations' benefits. Bradley Dewey, part owner of and investigator from the Dewey and Almy

Chemical Company, expressed his suspicions to TIIC that the reason one process had not been thoroughly investigated "is that most of the other fellows are afraid of the process stepping on some of their own pet secret processes."[83]

Interindustry rivalries might have led those sending investigators abroad to report less value in order to keep out competitors. Initial plans for FIAT warned that "the success of FIAT will of course depend in part upon the extent to which it is able to function exclusively in the national interest (during the SHAEF period in the national interest of the United States and of Great Britain), as ever against the interest of any particular individuals or business concerns."[84] Despite that, the decisions about which firms would be asked to send investigators were made by industry-specific panels, manned by representatives from particular companies. The Communication Subcommittee of TIIC took the extra step of checking, "through attorneys, that we are not under restriction to be sure that Panels are a fair representation of industry," and members battled over whether to include particular companies, including large ones such as Westinghouse (one member was strongly against it for unstated reasons).[85]

In terms of chauvinism, at one exhibition of German food-related technologies held in Atlantic City, discussion became "highly explosive," with one "impressive looking individual" attacking German cans as being inferior to products from his native Norway and another man—later found to be associated with a US canning company—hung around the exhibit all day, loudly insisting to other visitors that "American industry will always produce tinplate cans cheaper than the cans on display can be produced."[86] The National Machine Tool Builders' Association expressed a strong desire to have German machines shipped to the United States for study but was concerned about plans to display these confiscated German machine tools in public museums, because "they might get entirely too much attention and create in the minds of the average citizen the idea that American machine tools are not as good as German."[87] There was a clear concern about maintaining a consumer impression that American-made products were superior, whatever the in-house assessments and interest in German techniques.

Re-evaluating the Value Question: What America Gained

Accurately measuring the dollar value of technological exploitation is impossible. Science and technology might sometimes fall under the umbrella of the term "intellectual property," but intellectual property is not the same as physical property. A machine tool taken as reparations has a certain value that is possible to estimate in terms of market rates, including depreciation and obsolescence. Its removal or destruction is a loss to German industry (though if obsolete, its replacement by newer technology might be to the long-term benefit of the economy). If it is utilized in America, it is to the benefit of American industry (assuming its value is greater than the nontrivial transaction costs of seizure, disassembly, shipping, reassembly, and getting it into productive use). Intellectual property defies even this "easy" accounting. Investigating a technology or process robs the inventor only of the exclusivity of the knowledge. The technology loses no value for its inventor unless and until it spawns direct competition.

Depending on how we choose to count, we could estimate anything from enormous benefits to the United States to a net loss. For a higher number, we could add together the research and development costs German firms invested into the technologies that American forces investigated, the training and recruitment costs for scientists brought to the United States, and the increased market value of industries that investigated German technology (controlling against those that did not). This would certainly capture the hopes and expectations of those behind these exploitation programs, but it would fundamentally misrepresent what American industry gained and German industry lost.

It is important not to go too far in disparaging the economic value of the FIAT investigations. They produced real improvements to some American industrial technologies, even taking into account the disappointments and difficulties discussed previously. Speaking with German technicians and studying Germany technology sometimes inspired American engineers, even when the German technology was not any particular improvement on American techniques. When TIIC requested feedback from companies about their experiences in Germany, this cross-fertilization of ideas came up in several of the responses. DuPont Chemicals noted that though there were few direct gains, "the important thing is that we have obtained the basic ideas for further development in engineering from this Technical Mission."[88] Sidney Kirkpatrick, editor of *Chemical and Metallurgical Engineering*, admitted that a "white carbon black" made in America had used completely different processes than found in Germany but credited the German development for "at least planting the seed of the idea."[89] The American Smelting and Refining Company developed a product "not at all related to the German process" but credited the German example with giving them "confidence to embark on the project."

The company went on to say: "I believe you will find this type of thing is the most common benefit obtained from the TIIC investigations."[90] Several other reports mention "stirring the imagination" in similar ways.[91]

If we hope to assess the overall impact of these programs, we need to take into account costs as well. Sending investigators to Germany was not free. Each team required air transport, housing, vehicles, rations, and protection, all precious in a war zone. Time abroad also meant time not inventing and innovating at home. Investigations also cost diplomatic capital, goodwill, and legitimacy among the Germans being occupied and (the occupiers hoped) rehabilitated into democratic, capitalist, capable allies. The following chapters, especially regarding the French and Soviet cases, detail how far American eagerness to investigate Germany strained relations among allies.

The most important legacies of American investigations in Germany were not in accounting books but in American attitudes toward using its own science and technology to influence the Cold War world. American policymakers' faith in US "know-how" became a key factor in a number of broader Cold War strategies. It became a basis for Marshall Plan programs to raise Western European (including British) productivity so that these capitalist nations could raise standards of living and neutralize communist parties' appeal. The idea that the United States could improve standard of living in non-European, "third world" countries by sharing technical know-how—rather than direct capital transfers—allowed compromise between budget hawks and foreign aid advocates in Congress. The world's leading technological power could share that knowledge, the theory went, rather than directly sharing wealth.

Along with a sense of American technological superiority came a political will to crack down on illicit technology transfer. The United States has a long history of copying industrial technology from other nations, including the famous case of Francis Cabot Lowell recreating secret British textile machinery in America at the end of the eighteenth century. After the Second World War, however, US policymakers saw themselves as the guardians of the world's best industrial secrets rather than licensees or up-and-coming competitors. Intellectual property law grew much stronger in the postwar decades, and American technology licensing to others increased substantially. Diplomatic efforts of the United States even began pushing other nations to adopt American standards for business law, including licensing, intellectual property, and antitrust law.[92] This, too, tied directly into Cold War diplomacy, as fears of communist industrial espionage led to the attempted embargo of high-tech goods to Soviet-controlled countries via the Coordinating Committee for Multilateral Export Controls. The FIAT investigations are certainly not the only causes for this shift, but they played an important role.

In August 1951, John Green of the OTS was a guest on *The Eleanor Roosevelt Show* to discuss the National Inventors Council. Mrs. Roosevelt at one point asked Green if he thought important inventions would emerge from other nations, to which Green replied that he did: "We have a pardonable feeling that we have a monopoly on brains, but of course it isn't so. And there are some marvelous individual thinkers in Europe who are hard at work today and we can always hope that we will be able to borrow the knowledge of Europe, and I sometimes think of that as a sort of a reciprocal Marshall Plan to be able to take ideas of Europe." Roosevelt responded with a telling remark on Germany's new status: "Yes in the old days we did, once upon a time think of Germany—when they were allowed to think before the Hitler days [*laughs*]—of them as very good scientific research people and inventors."[93]

One quote hardly proves a general sentiment, but Mrs. Roosevelt was not the only commentator convinced by the 1950s that the Germans were not quite such a threat as they once had been. In April 1950, one author felt the need to write about "American chemists . . . not fully appreciat[ing] the amount of fine work that is being carried on in Europe today" in chemistry and recommended more people review the *FIAT Review of German Science* for details.[94] A 1953 article in the trade journal *Chemical and Engineering News* dubbed America's technological edge "our Maginot Line," suggesting that a public poll would show Americans saw the nation's security as based in (1) the atomic bomb and (2) "the great American Production Know-How." (The intended connotation here seems to be of the Maginot Line as an impenetrable wall rather than a security system quickly bypassed by German technical ability.)

An enormous body of writing debating how far it makes sense to talk about the "Americanization" of Western Europe (or the world) after the Second World War has emerged. As historian Jonathan Zeitlin writes, introducing a compendium of such essays in *Americanization and Its Limits: Reworking US Technologies in Post-War Europe and Japan*, "Few historiographical propositions are more deeply entrenched than the claim that the transfer of US technology and managerial know-how lay at the heart of the extraordinary economic growth experienced by Western Europe and Japan during the 'golden age' of the long post-war boom."

Science became, as Ron Doel has argued, "a vehicle to promote American values and interests in the post-war world."[95] John Krige, in turn, has expanded on this theme in his influential *American Hegemony and the Postwar Reconstruction of Science in Europe*.[96] There, he argues that American policymakers (both in government and in private institutions such as the Ford and Rockefeller Foundations) sought to use grant funding and scientific cooperation to reconfigure European science to more readily follow American models, and thereby to recruit these elites into a more pro-American mind-set.

Ultimately, the investigations of German science and technology by T-Forces, CIOS, FIAT, TIIC, Project Overcast/Paperclip, Alsos, the Technical Oil Mission, Army Ordnance, Naval Intelligence, the Strategic Bombing Survey, various unofficial or local groups, and the slew of British, French, Soviet, and other countries' teams who would later share reports, all likely had a modest impact on America's economy. This is far from saying they were unimportant, however. Though it has not been a focus here, there are still the moral issues involved in bringing scientists into the United States, some of whom had been ardent Nazis.[97] A few very clearly committed war crimes, such as actively participating in using (and even executing) enslaved workers or running inhumane experiments on human subjects.

One company, Caducean Press, brought advertising panache to its resale of a specific set of FIAT records: the results of the Nazi medical experiments, including those for which the researchers had been sentenced at the Nuremberg war crimes trials just a year prior. The firm advertised records "long kept secret," such as: "From Himmler's cave on a hillside near Dachau, records of physiological experiments on inmates of concentration camps. . ." and "From the laboratories of the Kaiser Wilhelm Institute, pathological anatomy of rare conditions found in the brains of mental patients who were executed. . ." For a "nominal cost" for translation and distribution, Caducean would send these files to customers. Those whose consciences might have stirred were assured that "the director of one of America's great research centers" had endorsed the idea of America learning from Germany's medicine.[98]

These intellectual reparations programs had important direct and indirect effects on America's industrial policy, postwar economic planning, diplomacy, and intelligence community. The lasting legacy of FIAT is not that it likely granted America many billions of dollars of value in technology. It is that before FIAT, FIAT seemed like a tremendously good idea to a huge array of American businessmen and policymakers; afterward, it seems almost silly to have bothered. This lesson in the difficulty of technology transfer was not unique to America, but, as we will see, it played out very differently in each of the Allied powers occupying Germany.

2

British Scientific Exploitation and the Allure of German Know-How

As the Second World War moved toward its end, British policymakers saw in front of them a tremendously different world than did their American cousins. In contrast to American worry about how to make use of all that excess American industry and lead the "free world" against the Soviets, the Houses of Parliament worried about enormous economic challenges and the threat of the United Kingdom becoming a second-class power. This was especially galling for those who could remember when the British navy was feared the world over, protecting a global empire and London's position at the center of international finance. Leading into both world wars, popular literature at home had warned of the threat of "Hunnic" hordes invading the British Isles, and now terrifying new weapons had smashed into London. Preventing this from happening again was, to say the least, a priority.

On the economic front, the United Kingdom had gone deeply into debt, especially to the United States, in order to pay for the war. Paying down that debt required foreign currency, meaning exports. For decades, the Commonwealth nations of Australia, Canada, India, New Zealand, and South Africa had been a captive market, but British policymakers dedicated to retaining the empire faced new opposition. India, in particular, had been an important market but had been promised greater postwar independence in order to secure peace and the contribution of tens of thousands of Indian soldiers to the Allied cause. American leadership—now in a stronger diplomatic position as a major creditor—was also generally hostile to empire. Furthermore, the United Kingdom had been forced to sell most of its foreign investments at any price in the 1930s and 1940s to buy weapons and supplies. Britain would need new and expanded export industries, then, while having less access to cheap, raw materials than before.

There was one resource that British policymakers saw as a potential savior: the "unsurpassed . . . genius" of British inventors.[1] British inventions and innovations had been absolutely crucial in the Allied victory. The key example was radar, which was one of the most decisive developments in the war, though British contributions to atomic weapons and aircraft design were also critical. Perhaps, some lawmakers thought, British scientists and technologists could invent products and develop new industries, guided in part by German research.

Simply having the "best" ideas is not the same as having competitive industries, however, and British policymakers saw their recent history as one of British genius being co-opted by others. Lord Riverdale, chairman of the Advisory Council of the Department of Scientific and Industrial Research, expressed a common sentiment in March 1944: "It is very easy to point out a dozen or more first-class inventions that have been invented in this country. Nobody would take any interest in them and they have been bought by the Germans and either used as they were or applied to some research which they were doing and for which they have afterwards obtained very substantial results."[2] Applying their "Teutonic genius" for applied engineering, the Germans had built powerful cartels in chemicals and other fields while British industry dawdled.

British policymakers, then, had somewhat different questions on their minds as they faced the political, logistical, and financial challenges of occupying a section of defeated Germany. What policy mechanisms could halt a perceived ongoing decline in British industry and empire? More specifically and urgently, how could they boost exports? Could this Germanic engineering capability be somehow brought to bear for British benefit in the longer term? Could these efforts to take German industrial science and technology somehow be woven into an even greater priority—that of building a closer relationship with the United States? Debates about these issues filled the halls of Parliament in the 1940s and 1950s.

British efforts to take German science and technology were fundamentally shaped by a growing realization that taking technology is not as simple as copying documents and writing reports, because these reports cannot capture the "know-how" component. Know-how (also today sometimes called "tacit knowledge") describes the skills and knowledge gained through hands-on experience that is difficult or impossible to write down. A classic example is how to ride a bike, which you would never learn from a textbook alone. In the context of industrial technology, a report might capture the chemical formula for a dye or a patent filing, which is useful information but far from all you need to reproduce the chemical. Experience, trial and error, and this overall know-how component is extremely valuable, even when the individual facets are too minor to patent, and usually the only way to acquire know-how is through in-person training. In this way, British efforts were shaped by a growing awareness that a British firm might be completely unable to effectively and efficiently reproduce a technology through FIAT/BIOS reports, no matter how well that report is written.

As British policymakers and industrialists became disenchanted with FIAT/BIOS report writing, they turned toward what they saw as a practical, longer-term tactic for turning German science and technology to the benefit of British industry: intellectual property law.[3] To British planners, it was simply naive to think they would occupy Germany forever and could keep its potential for military resurgence ground into the dirt. In their understanding of British-German economic history, though, Germany had not needed a military occupation to steal and profit from British invention in recent decades—they had accomplished that just fine through licensing agreements, exploiting differences in patent systems, and building powerful cartels. In this context, debates about whether and how to reform domestic patent law became connected to how to reinstitute patent protection in the British zone of occupation in Germany. In this effort to structure a permanent "brain drain" from Germany to the United Kingdom and its colonies, intellectual property reform in the United Kingdom both influenced and was influenced by diplomatic, political, and economic developments in the British zone of occupation (and later the Bizone and West Germany).

The United Kingdom Enters the Race for German Industrial Science

Given the obvious parallels between Britain's position after the First and Second World Wars, it is worth briefly recounting British-German relations in the interwar period. The 1920s to 1930s, after all, was the most obvious model policymakers had for how to act in the 1940s. In particular, much of the story of Anglo-German relations in these decades revolved around questions of reparations and war debts, and both positive and negative lessons from this period were fundamental in policy toward Germany after the Second World War.[4]

Britain had borrowed enormous sums from the United States to finance the First World War—even in 1934, the country owed 4.4 billion USD, or about 150 percent of its gross domestic product, after paying 44 percent of government expenditures toward the debt in the mid-1920s. Meanwhile, though Britain had pushed back against even harsher French demands, the German interwar government had agreed to pay a similarly staggering sum in reparations. As the German economy collapsed, the United States stepped in to loan even more money to Germany, who paid Britain, who paid the United States—all leading up to the stock market crash setting off the Great Depression in the 1930s. This was not a situation that British policymakers were eager to repeat, given the opportunity.

In a wider lens, Britain's gradual eclipse by Germany (and the United States) as an industrial powerhouse over the late nineteenth and early twentieth centuries stoked British fears of relative decline, and this mind-set, too, survived to shape British scientific exploitation efforts.

To repeat a bit of the wartime infrastructure described in chapter 1, the Allied armies combined operational structures into the Supreme Headquarters, Allied Expeditionary Force (SHAEF). Within SHAEF's intelligence division, dubbed "G-2," a joint US–UK agency called the Combined Intelligence Objectives Subcommittee (CIOS) focused on identifying targets for intelligence missions that might be useful in preparing for the upcoming shift to the Pacific theater.

With the dissolution of SHAEF in July 1945, CIOS split into American and British components. The latter of these, now called the British Intelligence Objectives Subcommittee (BIOS), switched from being a fully military intelligence program to serving under the direction of the Board of Trade.[5] It continued to receive logistical support from the so-called T-Force units within the military, who were responsible for racing along the front lines across Europe, seizing and guarding scientific and technical targets to prevent looting or sabotage.

In an independent but related effort also sponsored by the Board of Trade, a panel chaired by Sir Charles Darwin—descendent of the famous author of *On the Origin of Species*—orchestrated an effort to recruit German scientific and technical personnel for the benefit of British industry. This "Darwin Panel" mirrors the well-known American Operation Paperclip in some ways, though the Darwin Panel focused primarily on civilian export industries and Paperclip primarily on military aerospace technology.

The Board of Trade set out to recruit investigators from industry for both

TABLE 2.1.
Comparing US and British components of CIOS

BIOS committee members	TIIC (US) committee members
Foreign Office	Department of State
Naval Intelligence	Office of Naval Intelligence
Military Intelligence	Intelligence Div. (G-2) of War Dept. General Staff
Air Intelligence	Army Air Forces Intelligence
Ministry of Supply	Foreign Economic Administration
Ministry of Economic Warfare	Office of Strategic Services
Ministry of Aircraft Production	Office of Scientific Research and Development

programs. These investigators would receive military uniforms, ceremonial rank, housing, and transportation (arranged by the T-Force units) to investigate German targets identified by CIOS and BIOS, after which they would write reports for use by the rest of their industries. The Board of Trade and BIOS expanded their objectives in the late 1940s from strictly military targets (e.g., plants that manufactured V-2 missiles or scientific research facilities) to include targets useful for civilian industry (e.g., furniture factories). They cooperated with the United States' Field Information Agency, Technical (FIAT), which had essentially taken over for the American half of CIOS. A direct liaison group dubbed FIAT (Britain) (or FIAT [BR]) mirrored FIAT (US), though BIOS remained the primary agency in charge of British efforts.

Within the British occupation zone in Germany, the Control Commission for Germany, British Element (CCG/BE), initially assisted but later resisted these direct exploitation efforts. I discuss the rationale for this shift later, but essentially these military governors saw industrial exploitation as contrary to their mandate to rebuild the zone's economy and thereby reduce the cost of occupation. The major players on the British side, then, were BIOS, in charge of overall exploitation; FIAT (BR), as liaison to the United States and later to France; the Darwin Panel, independently focusing on hiring German technical personnel; the Board of Trade, as the government agency in charge of overall policy and industry liaison; and the CCG/BE in charge of running the German zone. The intelligence units of the British military branches orchestrated their own focused investigations, but these were not on the same scale as other programs.

The British scientific intelligence efforts lacked an exact equivalent of the US Department of Commerce's Publications Board. Instead, the Board of Trade advertised CIOS, BIOS, and shared FIAT (US) reports via its newsletter and directly to industry magazines and trade associations. Her Majesty's Stationery Office filled orders for copies of these reports. Extensive publicity campaigns aimed at reaching even "smaller firms—and it is probable that individually and collectively they have the most to benefit from this insight into German methods—[who] will not use the material unless it is brought pretty forcibly to their notice."[6] The first copies of BIOS reports arrived in industrial cities throughout the United Kingdom in late 1946, a policy that led to requests for inclusion from county libraries miffed at being left out as well as requests for fewer copies from city libraries lacking both demand and shelf space. A small exhibit of reports and prototypes in Bristol, bulletins in the *Board of Trade Journal*, and occasional press releases supplemented efforts to advertise BIOS reports to all potentially interested parties.

Initial Plans Hit a Snag: The Written Word Is Just Not Enough

Initial British proposals for exploiting German technology for civilian industry emphasized that everyone should benefit—not just big, well-connected firms. As a result, they emphasized duplicable, written reports. This decision to serve industries rather than firms, cast in terms of "fairness," avoiding "jealousy" and favoritism, and making the "ethical" choice, was built into the structure of BIOS and the Darwin Panel scheme.[7] Though investigative teams were felt to lose productivity beyond three to four members, they sometimes sprawled in order to "be fully representative of the industry concerned i.e. they must include representatives of the main Trade Associations and the main <u>NON</u>-Association firms."[8] Whenever possible, competing firms were placed on the same team, with the anticipated result of each holding the other accountable for including everything in the final reports.[9] Official policy discriminated against providing aid to firms that "refused to take part in BIOS investigations for fear of letting in their competitors . . . however much common humanity may lead us to sympathize with their attitude."[10] Written reports on German technical processes flowed through the country, and BIOS officials anticipated significant economic value to flow with them.[11]

Reception for these reports was not entirely positive, however. Within months of the creation of BIOS, complaints from end users grew in volume and urgency, saying that reports of any kind were insufficient. The first formal meeting of the Darwin Panel in December 1945 addressed this issue straight away. The chairman, Sir Charles Darwin, "agreed that it was far better actually to employ Germans in industries where the full power of their experience and criticism could be brought to bear, then to interrogate them. This was the only method of discovering the use of the people whom the Panel was considering."[12] Bringing these scientists and technicians to the United Kingdom was a necessity for control purposes, he felt, as their knowledge and expertise would otherwise live on despite any industrial dismantling—the knowledge lived in the people, not the equipment, data, or patents.

In July 1946, several businesses who had contributed investigators began lobbying for the opportunity to send their investigators back to Germany on follow-up visits. In a Board of Trade meeting to discuss this proposal, a representative of the Ministry of Fuel and Power argued that "BIOS reports are valuable up to a point, but for firms seeking to copy a machine or introduce a process developed by the Germans a further and more detailed examination is almost certain to be essential. . . . BIOS Reports vary greatly in their practical value to industrialists and . . . few, if any, are likely to provide adequate information . . . to introduce and develop a German process in this country."[13] A Board of Trade representative agreed: "The information contained in BIOS reports . . . is quite insufficient to permit potential new users, particularly those with limited research facilities, to set up and operate the process."[14] A report from October 1946 added that "experience has shown that if industry was left to prepare the reports they were of little value to firms which had not taken part."[15]

These might seem to be complaints about the *quality* of the reports, rather than about written reports in general, but leaders from industry and BIOS-related agencies were quite clear that their problem was not with the detail or prose. In fact, both British businessmen and their American counterparts agreed that British reports were the best to be had. Derek Wood, head of BIOS, boasted in October 1946 that "BIOS reports are widely recognized as being superior to those produced by the Americans. Our system of putting competing interests in the same team has undoubtedly done much to prevent concealment of the really interesting topics. . . . Industry has lived up to its side of the deal, firms and associations sparing neither trouble nor expense to make the reports comprehensive and instructive."[16] John Green of the US Department of Commerce's Office of Technical Services admitted that he was "envious of the polished materials you make available."[17] The French were no threat, either, as they would not or could not produce reports even for their own industry. (In British eyes, this French failure was due to "lack of organization, personnel and equipment," which they saw as characteristically French.)[18] The Soviet Union, too, cast a vote of confidence for British reports by means of purchasing every one at a cost estimated to be more than 400,000 USD per year.[19]

Complaints about the limited utility of BIOS reports reflected a conscious, ongoing struggle with the difficulty of capturing technology in written form. Industrial firms and trade associations pushed aggressively for finding methods of transferring tacit knowledge. At various points, they requested on-site inspections of German plants. They wanted to embed their engineers in these plants for weeks or months, hire German technicians, and end international cooperation in scientific exploitation, since that threatened to erode any advantages gained. Textile and chemical company Courtaulds wrote to BIOS in 1946, requesting additional inspections of IG Farben's plants, as even after sending a team, the information necessary for building a new facility "can only be obtained from the Dormagen technicians."[20] The Association of British Chemical Manufacturers got quite heated in their demands for follow-up investigations. Such reports, as the writer of an October 1946 letter argues, were very rarely sufficient to transfer a technology or process, and "we have not spent all the time and trouble in organizing investigating teams merely to produce a row of reports on the shelf. . . . First hand investigation would eliminate a great deal of the usual trial and error in setting up a plant here. . . . Much of the 'know how' is impossible to put into words."[21] "In practice . . . no amount of 'given' information can ever be a substitute for the information obtained in the hard school of practical experience. . . . The arguments above seem to be so conclusive that there can be no reply."[22] Similar statements from other industries appealed for change in the Darwin Panel and BIOS agencies.

These complaints were successful. Starting in July 1946, investigators received permission to make longer follow-up trips to German facilities, sometimes without the balanced teams of competing firms and rarely requiring extensive reports. At a BIOS meeting in mid-1946, one officer expressed ongoing concern about "abusing" their role as occupier by aiding individual British firms. He emphasized that the Darwin Panel had only ever been agreed to with assurances that whole industries, rather than individual firms, would benefit. Still, this principle being accepted, there was "general agreement" among those present that BIOS was "not really of any general benefit" anyway, as the firms sponsoring investigators received almost exclusive advantage.[23] "In all honesty," according to a later report, "BIOS investigations are . . . to some extent equally discriminatory in favouring firms represented in teams as opposed to firms who have to read reports."[24] "The practicable advantages of the Scheme were set off against the criticisms . . . of the discrimination to be shown to the favoured few," and despite fears that "it was wrong in principle that a specific firm should be able to acquire . . . trade secrets which were not for sale," the panel approved the scheme, adding only a stipulation that German firms receive some pay for their trouble.[25]

This shift did not go unnoticed within Germany, and orders to benefit specific firms upset even some British officials. Decades later, when recounting his wartime experiences, one British T-Force official recalled a case that had especially frustrated him at the time: A civilian, Mr. H. L. Muschamp, visited Germany as an investigator in the textile manufacturing industry. While there, he began ordering the logistics "T-Force" unit to ship valuable machine tools to H. L. Muschamp, Ltd. The officer in charge objected, pointing out that T-Force was a military force meant to serve the broader public good and could only ship to official government agencies. Two days later, "T-Force HQ received a message from London: 'Consign the machinery tools to Ministry of Supply, c/o H.L. Muschamp Ltd.' "[26]

Sharing Know-How: A Practical and Diplomatic Problem

The British, French, and American technical exploitation programs all deeply influenced one another, both directly and indirectly. Officials in all three nations used the threat that the others were already taking German scientists as a key argument for building and expanding their own programs.[27] From the British perspective, CIOS had been one building block of the Anglo-American "Special Relationship" in intelligence sharing that had blossomed during the war, and BIOS-related programs were initially another avenue for binding the nations more closely together. Intelligence in general was an area in which the British were still the senior partner in the relationship, tutoring the brand-new Office of Strategic Services on everything from tradecraft to analysis methodology.[28] In the realm of military technology, the nations worked together closely on sweeping up every researcher involved in German aeronautics to keep them away from the Russians.[29] Some tension arose when the British military pushed for more of these scientists than the Americans wanted to give, but the British planners were careful to preserve goodwill. Sometimes this meant forfeiting scientists they wanted; other times, this meant hurrying things along, with concerns that delays were "likely to cause unfavourable repercussions with [the] Americans, who under agreement are supposed to receive results of research work done by British in UK and who have renounced their claims to Germans in several cases on [the] understanding [that the] British would be taking them."[30] Diplomacy and exploitation were closely linked.

The perceived need to shift to a tacit knowledge focus, then, came at a price, because it split with American policies and expectations. Written reports could be shared with allies. Personnel with hands-on experience, and British engineers implanted in German factories, could not. The British decision to deprioritize written reports in favor of know-how, made in this international context, is even more striking a demonstration of the elevated importance of science- and technology-based exports for the postwar state.

Cooperation or Secrecy?

Cooperating with the Americans during the combined command phase of military operations was a given, especially when it came to finding anything that might spare Allied lives in the Pacific Theater. When it came to investigations of industrial technology, though, British policymakers seriously debated whether to invite the Americans to take part, or simply to go it alone. There was a general feeling that industrial exploitation was a "natural extension" of the military scheme, but that did not necessarily mean that there was any "moral obligation" to cooperate in this area, too.[31] The stated purpose for the third meeting of the Darwin Panel was to consider *whether*—not *when* or *how*—lists of German scientists required for employment in the United Kingdom should be exchanged with the Americans.[32] The matter "was settled for defense," but there were "fundamental differences of outlook held by civil industry and several additional difficulties," among them "whether, if cooperation was decided on as a policy between Governments, the American Government was capable of supervising adequately the activities of big business."[33] This concern about American decisions being very heavily influenced by business interests was a theme throughout internal debates on cooperation and coordination with the United States.

In the end, the panel voted six to four in favor of sharing the information fully: "those concerned with Trade Departments voting against, and those who were voting on general principles voting for the motion."[34] The Board of Trade concurred later that month that "the balance of advantage undoubtedly lay in full co-operation," precisely because (at this relatively early stage, before emphasis shifted to implanting investigators to acquire know-how) the value to be gained by reading American reports was assumed to be tremendous. Once the decision was made, the Board of Trade started planning ways to circumvent any political opposition by leaking rumors to American industry that the United Kingdom was already benefiting in major ways.[35] Big business, they felt, usually had its way in America. If business leaders saw the possibilities, so would US politicians.

Meanwhile, the Darwin Panel threatened British doubters with the possibility of the United Kingdom yet again missing an opportunity to lead in industrial technology, playing on long-standing anxiety about British industrial "decline." They also hinted that the Americans, French, and Soviets were already in a race for German science. Early coordinated efforts went smoothly. The American and British exploitation agencies cooperated broadly, mostly rubber-stamping requests for each other's investigative teams to cross the lines dividing the Bizone (as the American and British zones of occupation in Germany were nicknamed).

This competition with the Americans—to some degree a reality, though exaggerated when compared with actual American planning documents— drove British policy. The participants of early meetings of the Darwin Panel spent much of their time worrying if the contracts they were offering German scientists were "at least as favourable as the Americans were alleged to be giving," and if they were working quickly enough, as "speed was the essential factor since the Americans were approaching these people with good offers."[36] This fear that the Americans would hire all the Germans worth hiring persisted. In July 1946, the Board of Trade considered scaling down investigations substantially in favor of escalating the direct, long-term hiring of German personnel. Proponents argued that "private American businessmen were active in their zone," undertaking "private negotiations of the kind envisaged," thus "the Americans must have found some means of paying the Germans for their technical services" despite the bipartite agreements to the contrary (in fact, they did not).[37] The board instructed the British Embassy in Washington to feel out the United States on the idea of paying German scientists but instructed them that by no means should they let the Americans know that BIOS intended to do so.[38]

Even the CCG/BE, the day-to-day governing agency for the British zone of occupation, which generally opposed the exploitation efforts, exhibited the twin fear of either displeasing or losing out to the Americans. In late 1946, following months of backroom suggestion that BIOS-like agencies might be hindering the rehabilitation of Germans into productive, democratic citizens, the CCG/BE asserted more strongly that "continued piracy of German methods" must wind down in favor of building up the German economy.[39] Even in this communiqué, however, they admitted that it was impossible to cease operations so long as the United States continued theirs, as this would give America a monopoly on hiring. The end date must be coordinated bilaterally—a decision that effectively extended the life of both American and British programs by some months, as each wanted to be sure to at least match the other.

This dynamic of relative cooperation began fraying as know-how became a focus for the United Kingdom and written reports—especially those issued by FIAT (US)—fell in the Board of Trade's esteem. American bibliographies were "useless," as they contained too much information with insufficient depth and clarity.[40] If British firms were unlikely to truly acquire new technologies via FIAT reports, then purchasing reports even at cost was simply a waste of hard currency, and at a time when the Treasury needed every dollar. By mid-1946, as BIOS shifted under pressure from industry toward maximizing tacit knowledge, new proposals envisioned "a subsequent phase to which . . . the BIOS plans of equal participation rights to all United Kingdom and United States industries cannot be extended."[41] In September 1946, US Department of Commerce bibliographies of FIAT reports were officially "not to be made available to industry in this country . . . in view of their unsatisfactory character."[42] If British firms found out about US offerings and requested copies anyway, BIOS would fulfill these requests, but otherwise the costs were too high and the benefits too low.

A retrospective report from BIOS at the end of 1946 concluded that there was a "common belief . . . that the Americans are in most forms of exploitation always one jump ahead of us and that they invariably make the scale of our effort look small," but in reality the ten thousand British investigators dramatically outnumbered the approximately six hundred from the United States.[43] In early planning, having a large number of investigators had been a burden, only allowed in order to be fair and fully represent entire industries. Now this same feature was seen as a key benefit, and one that clearly could not be shared with anyone else: "it must inevitably have been much to our advantage, at this present time of reconversion to peace-time production, to have this vast number of technical men from our own factories walking round German plants getting first-hand knowledge of the methods of German industry."

"Your guiding principle," one memorandum ordered BIOS subdivisions in late 1946, "should be that a substantial 'bite out of the apple' is better than a 'smell all round.'"[44] This focus on in-depth "first-hand knowledge" rather than breadth of information led to "very vigorous . . . very critical" reactions to American decisions over the course of the BIOS-related programs. The diplomatic stakes here were not alliance-breaking by any means—despite their disagreements, both American and British representatives on a number of levels celebrated the continued good relations enjoyed by BIOS/FIAT, even ending their collaboration with a party in London (the costs of which led to complaints from Treasury). Still, the initial planning and ideal scenario was one in which both nations shared and shared alike, bringing their economies up to the cutting edge and enhancing the "special relationship" along the way. A shift to emphasizing long-term human contact (through implanting British investigators or importing German specialists), and thereby to attaining the know-how component of technologies, was no idle decision.

The Soviets were already largely in agreement with these principles, yet the tensions of the early Cold War meant that British diplomats publicly held to their faith in written reports. As early Cold War tensions escalated, an article in the Soviet state newspaper *Pravda* on March 21, 1947, accused the United States and United Kingdom of exploiting German technical "secrets" at the expense of their allies. The head of the UK delegation to the Council of Foreign Ministers (the quadripartite planning group that set cross-German policy on increasingly rare issues that achieved unanimous approval) immediately requested that the CCG/BE provide him with additional information on *Pravda*'s claims that "BIOS reports [were] valueless since the information they give is totally inadequate."[45] The official British response was to argue that these reports were perfectly adequate, and that the Soviet Union must think so, too. The Soviet Embassy routinely bought both American FIAT and British BIOS publications at a cost that the head of FIAT (US) estimated at 400,000 USD per year ("a fact which will not be forgotten when the time comes to consider the loan of money to Russia").[46] Internally, BIOS had long since come to a similar conclusion—and switched policies because of it.

Taking a Long View: Intellectual Property Law and the Occupation of Germany

Though aspects of the BIOS and Darwin Panel schemes operated through the early 1950s, most of their functions wound down from 1947 to 1948. Conventional accounts of the postwar technical exploitation, both in the American and British cases, wrap up with the story of the CCG/BE asserting the importance of normalizing the German economy in order to protect the zonal / West German economy from the "harm" of exploitation and spare taxpayers the costs of occupation. Historian John Gimbel describes the end of American efforts as a victory for a faction of "Governors" over "Exploiters."[47] John Farquharson rejects this framework as "not a valid summary of the British occupation" but does so largely because the situation was "by no means so one-sided that 'exploitation' is necessarily an appropriate word," considering the United Kingdom's investment in Germany, and that German firms could also purchase BIOS reports (and some, in fact, did so).[48] Certainly, the CCG/BE fits what Farquharson described as the "governors vs. exploiters" model to some degree, as it increasingly resisted direct exploitation efforts—to the frustration of some firms. Worried about the morality of British policy and "criticism that Germany was being exploited for the benefit of British industry to the exclusion of our Allies," its director suggested that British firms should pay the German firms they visited for their information.[49] The chairman of at least one BIOS group agreed in principle: "the time is coming when relations between firms in this country and those in Germany will have to be placed on a more normal footing if further useful information is to [be] obtained by individual firms."[50]

This framework of the CCG/BE as "governors" protecting the interests against the "exploiters" misses a fundamental motivation for CCG/BE's stance, however. While the CCG/BE was bureaucratically opposed to BIOS-like agencies, they were not actually opposed to exploitation of German science and technology for the benefit of British industry. Quite the opposite: they worked from the war's end through the 1950s to set up a patent and trademark system within Germany that would, among other goals, "open the door to a substantial flow of German ingenuity" into the United Kingdom, which could "hardly fail to be other than a benefit to the trade and industry" of the nation.[51] These policies were guided at least in part by the sudden perceived importance of "know-how," seen both in BIOS investigations and in parliamentary debates on how to reshape British patent law for the postwar world. The CCG/BE and other long-term planners were indeed interested in building Germany's economy and sparing taxpayer subsidies, but they were at least as interested in building a permanent harness for German minds.

British patent law was built on concerns of acquiring foreign technology and keeping British innovation in-house. "Letters of Protection" (later, "letters patent") under King Edward III in the fourteenth century encouraged foreign craftsmen to settle in England and teach apprentices, a practice renewed in the mid-sixteenth century. According to historian Christine MacLeod, "acquisition of superior Continental technology was the predominant motive for the issue of patents under the guidance of Elizabeth I's chief minister."[52]

This aim heavily influenced patent policy, including when it came to attracting German technicians with mining technologies in the 1560s. Eventually, other justifications for patents came to predominate legal and political rhetoric: an inventor's natural right to his invention and encouraging industry to invest in research and development. Still, while the earlier rationale of drawing in outside talent faded, it remained a viable rhetorical tool when British worries about falling behind their economic rivals reemerged in the twentieth century.

An early sign of the revitalization of this line of thinking came from the British chemical industry in the 1890s to early 1910s, as they successfully lobbied for the 1907 Patents and Designs Bill.[53] Responding to arguments that German chemical cartels were patenting a wide range products in the United Kingdom simply to keep British firms from developing competitive skills, the bill required that patented products be actively manufactured on British soil. If they were not, the patent would be invalidated. The bill was hotly contested in Parliament, but a key argument in its favor was bitterness over several technologies that had been invented in the United Kingdom but had only been developed and patented in Germany. Initially, the act seemed to serve its purpose of offering British employees opportunities to acquire German chemical know-how, as German firms quickly built small factories in the United Kingdom to retain their patents. However, the Board of Trade (which was in charge of the Patent Office) never seriously enforced the law, and these factories closed as soon as this became apparent.

The Second World War spurred invention and innovation, yet nearly every country suspended its patent and trademark systems. "Trading with the Enemy" statutes sprang up around the world, banning commerce with opposing nations. Both Allied and neutral nations confiscated German trademarks and patents, which were seen as "weapon[s] of economic penetration . . . of immense strategic value."[54] In the Americas alone, Ecuador, Nicaragua, and Venezuela each seized German trademarks between 1944 and 1947, while the United States placed German patents and trademarks under control of the Office of Alien Property Custodian.[55]

In July 1946, at an international conference in London, the governments of the United Kingdom, United States, France, Australia, Belgium, Canada, Czechoslovakia, Denmark, Luxembourg, Netherlands, Norway, and South Africa (but not the Soviet Union) came together to decide what to do with these seized German patents. In an agreement signed in July, they established that they would make all German patents open to royalty-free licensing, available to all, and that in return the Inter-Allied Reparations Agency would not count these seized patents against reparations accounts.[56] This conference also codified additional international support for the BIOS/FIAT programs:

> Subject to the statement of the position of the French and United Kingdom Delegations as set forth below, it is the view of all Delegations to the Conference that the programme now in operation for obtaining, analysing and publicly disseminating German technology and "know-how" has proved of great common benefit and should be continued. At the suggestion of other Delegations, the Delegates of France and the United States will urge their Governments to request the military occupation authorities in Germany to give early consideration to utilising in this programme, so far as practicable, trained technical personnel and physical equipment which any other country represented at the Conference is able to furnish.
>
> The Delegation of the United Kingdom, while sharing the view that the programme now in operation has proved of great common benefit, and declaring that the Government of the United Kingdom would continue its practice of publishing all information of this character received from Germany, was unable to participate in any recommendation on this matter because there had been no time for the consultation with the occupying authorities in Germany which the Government of the United Kingdom considered to be necessary.

In addition to copies of German patents filed in their home countries, the United States seized the German patent office itself. American teams microfilmed its contents and shipped copies to London. This created considerable ire from France and the Soviet Union, who were denied full access. Only Britain's "excellent relations" with the United States "accorded, unofficially, certain privileges," among them this access to the full body of German patents.[57]

In occupied Germany, not only were all old patents up for grabs, but there was no system for filing new patents. Until a uniform policy could be decided upon with quadripartite consent, each zone was free to create and enforce (or not) its own regulations for allowing trademarks, patents, and copyright. From 1945 to 1949, it was impossible for Germans (or anyone else) to register new patents or trademarks within Germany, or for Germans to patent abroad (other than refugees or others who might be declared "not Germans").[58]

Recreating Germany's patent and trademark system was not a high priority for most of the occupying powers. Both the Russians and the French saw a new German patent office as undercutting their opportunity to acquire German technology. The British Foreign Office attributed Soviet resistance to a new patent office as a combination of ideological opposition to capitalist ideas of intellectual property and as an intended bargaining chip to be used in unrelated quadripartite negotiations they actually cared about.[59] If a patent office were to be created, Soviet negotiators demanded that the USSR have veto power over each patent application, which the United Kingdom interpreted as an effort to build "a window into the mind of West German investors" by allowing Soviet technicians to read and then reject patent applications.[60] France resisted any centralized bureaucracy that could result in a strong German state, including a central patent office, and pushed for a new International Patent Office in Brussels to handle German (and other) patents. The British Foreign Office attributed this initiative to "French jealousy" of the prewar German patent office, as France's system had "always been something of a joke with the more advanced and patent-minded Nations."[61] Actually, France was busy exploiting German ingenuity in their own way. The French military occupation authority unilaterally opened a patent office in the French occupation zone in early 1946 where Germans could pay to register patents and trademarks, an opportunity many German investors and engineers eagerly utilized as a basis for future claims for patent priority internationally. By resisting any other legal patenting across Germany, France effectively drew German "invisible capital" from across the divided nation into its zone.[62]

In contrast, the CCG/BE saw legal protections for intellectual property as exactly what would allow better British exploitation of German science and industrial advances. For the CCG/BE, reestablishing some kind of patent office was "an obvious decision," necessary for constructing "a modern State" (in West Germany, at least, as a unified Germany became less and less likely).[63] Creating a legal framework for British industry to license or buy German intellectual property would "open the way to the flow of German inventive ingenuity into the UK for our benefit."[64] Even if the CCG/BE were in favor of BIOS, its mission statement only covered investigating *wartime* advances in science and technology. Anything invented before the war or after the armistice was technically off-limits, and while investigators often stretched (or wildly overstepped) this line in practice, BIOS was not a long-term solution.

Without patent protections, rumors circulated of Germans hiding inventions from the occupation authorities. This would not only prevent any sort of BIOS investigation but was in violation of Law No. 25, concerning the control of scientific research in the British zone of occupation. Law No. 25, discussed at length in chapter 5, essentially forbade all military research in Germany and required extensive paperwork to approve any nonmilitary science. Internally, the CCG/BE admitted that Law No. 25 was "not really necessary" and "largely designed to combat a danger which does not exist" (i.e., the threat of Germans developing deadly scientific weapons that strict Allied control could prevent). Still, the law at least kept control of scientific research out of the bureaucratic hands of intelligence agencies.[65] CCG/BE administrators with science backgrounds worried that British intelligence was "not in general staffed by men of a type who are able to maintain good relations with high-grade German scientists. . . . We regard the maintenance of such relations as a cardinal point of our policy." Finally, without a legal marketplace, technology would still flow along the black market. "Some inventions disclosed to CCG/BE in confidence [were] reported to be getting into the hands of British Industrialists" by less than scrupulous means. German inventors, either desperate for work and willing to give up their ideas for any opportunity or unwittingly passing along their concepts in the course of complaints about the lack of a patent system, would describe their inventions to British officials, who were passing them along to British firms.[66]

British policymakers concerned with using German ingenuity to benefit British industry increasingly militated for reinstatement of a formal intellectual property system. Yet to understand the policies they put in place, we must detour to parallel debates happening in Britain, as they rethought their own intellectual property system in the 1940s.

Patents and State Involvement in Innovation in the 1940s and 1950s

While the CCG/BE and British Foreign Office debated how to rebuild the (bi)zonal economy and aid British exports, lawmakers in Parliament struggled to rewrite Britain's patent laws. After years of debate, a series of substantial amendments to existing laws passed in 1949. These amendments intended to tackle a number of issues, but a few fundamental concerns unified them. By making patents stronger (covering longer periods and allowing them to be more easily defended), they would promote economically useful innovation. Given recent wartime inventions, they addressed a need to clarify national security considerations in patent law (e.g., whether to allow patents on classified technologies or for the military to invalidate patents in times of war to promote cheaper mass mobilization). Finally, they sought to reflect a larger role for the state in general in questions of science and innovation policy.

These were many of the same issues that concerned governors of the British zone of occupation, if almost inverted. They wanted to *promote* the flow of German ideas overseas, whereas policymakers in London hoped to keep British invention in the United Kingdom. Officials of the occupation government sought to make secrecy next to impossible in Germany, while Parliament sought to create new spaces for keeping secret any inventions with national security implications. The connections between intellectual property debates in the United Kingdom and in Germany are often indirect—though sometimes they were quite explicit, such as with laws indemnifying those using BIOS reports from any later infringement claims—but the same parliamentary debates and policy discussions shaped both. These discussions became remarkably broad, challenging members of Parliament and British zone occupation government to reconsider the fundamental relationships among science, invention, and state intervention.

British patent law had serious problems in the interwar period, and these problems combined with popular anxiety about British "decline" and a looming German menace to inspire change to these patent policies. As one commentator put it in 1945: "The British patent system has few friends. There appears to be a general lack of confidence amongst the public (consumers), the inventors, the manufacturers and the investors in its efficacy and future operation."[67] The Right Honorable Sir Lionel Heald, attorney general in Churchill's cabinet, looked back at the prewar years from 1952 as another era: "With no easily assignable cause, there was a distinct sense of hostility against patents. . . . Patent litigation before the war had come to be too much associated with technical subtleties and ingenious scientific hypotheses and undue reliance was placed on the argumentative evidence of 'court experts.' This coincided with extreme specialization at the Bar and a consequent tendency to ignore or at least to attach little weight to general legal principles or the rules of evidence. The inevitable consequence was to stimulate judicial suspicion of the whole 'patents racket.' "[68] Even Bernard Shaw joined in the ribbing, asking why British patents offered so little protection even compared to the copyright that protected his plays.[69]

In contrast, the German patent office was a model for both American and British patent reformers in the early postwar years. Before the war, it had been effective and well-staffed, with 1,600 employees working in a seven-story-high, nine-hundred-room building that stretched over a six-acre estate. Its examiners were notoriously careful, meaning few patents were overturned by courts. More than a few articles written by both American and British patent lawyers and policymakers in this era note that French patents, in contrast, were notoriously poorly managed—even to the point that perhaps the reorganization of the French patent office was at least one small silver lining in the Nazi occupation.[70]

German firms' use of patents for cartel organization and monopolistic business practices around the world were, in contrast, seen as a corrupt threat that had worked against British interests for decades. In 1943, the House of Lords debated how to combat the threats outlined in the popular, hyperbolic exposé of Germany's "infiltration" of American industry: *Germany's Master Plan: The Story of Industrial Offensive.*[71] In 1944, the Ministry of Economic Warfare warned that there was "considerable evidence of German infiltration into Spanish industry" taking the form of the Germans "making available processing and patent rights and supplying technical plant, advisers and engineers." (It was, however, "not possible to assess with exactitude the degree of control obtained by such methods.")[72] This complex legacy would be—depending on perspective—a promise or threat for those managing Germany in the postwar years.

The president of the Board of Trade appointed a committee in April 1944, headed by Sir Kenneth Swan, "to consider and report whether any, and if so what, changes are desirable in the Patents and Designs Acts, and in the practice of the Patent Office and the Courts in relation to matters arising therefrom."[73] This committee would eventually issue three separate reports after eighty-one all-day sittings. These hearings stretched out over years, with reports issued in 1945, 1946, and 1947. The final Patents and Designs Bill of 1949 instituted most of these recommendations, which became the core of UK patent law through the 1970s, when international treaties and global economic shifts demanded new policies.[74]

The perceived threat of British industrial "decline" hung over the issue. An article bemoaning the state of British patents written in October 1944 noted that "during the eighteenth century we were the most technical nation in the world; during the nineteenth century by force of circumstances we became the financial centre of the world. It seems as if from 1900 to 1939 progress and finance in this country failed to continue their partnership."[75]

In both the Swan Committee's deliberations and in parliamentary debates regarding the bills emerging from the committee's reports, the speakers found themselves asked very basic questions about the interactions of science, technology, and society. Wartime experience had shown that the government could—and perhaps must—sponsor important research. Radar had been vital in defending against Nazi air raids, and in a nuclear age the importance of science for national defense was clear. What could a state do to sponsor technology effectively, though, without undercutting business? How could the state keep defense-related technologies in the country and ensure that international competition did not leave Britain once again at the mercy of an enemy nation who controlled key industries (e.g., chemicals)?

These were complex questions, and members of Parliament tackled them from a variety of angles. "If we are to [succeed in] developing research," a member of the House of Commons argued in 1944, "we shall have . . . to reconsider the whole question of our patent laws." Debate on patent law amendments led the House to spend much of the day debating such broad conceptual questions as "What is 'research'?," "What does the speaker mean by a 'scientific mentality'?," "Is economics not a science?," and "What would it take to be a 'scientific nation'?"[76] In all, the House spent for the first time in its members' memories (and perhaps ever) "a full day's Debate . . . devoted to the subject of research and scientific knowledge."[77] Some even saw it as a racial issue, with "the ingenious barbarians" of Germany and the "very good imitators" of Japan having particular advantages despite the "unsurpassed genius" of the British people.[78]

BIOS investigations had granted "three years of very great advantage" by revealing the "secret processes and prototype machines of the whole of German industry."[79] Yet this was not enough. Noting America's inability to use patents seized in the First World War to build an independent chemical industry, Viscount Maugham remarked on the questionable utility of patents and added that if Britain's seized chemical patents were of similarly marginal value, "something ought to be done about it."[80] He did not specify precisely what remedy might be possible, but German patents, at least, were not to be trusted to transfer knowledge without accompanying expertise.

This emphasis on a know-how component being key to patent reform resurfaced in a number of places. Several members thought the nation excelled at producing "first-class brains" at its universities but needed to focus on producing more "second-class brains" who had the technical experience and adaptable hands-on skills to turn inventions into exports.[81] Returning to American comparisons, Viscount Swinton reported in the House of Lords that in the United States, patents were "rather falling into desuetude," as companies preferred to "rely on being first in the field and having the know-how" (Lord Strabolgi replied that this was only because "in the United States the only thing that matters in the protection of patent rights is 10,000,000 dollars").[82] Lord Chorley advised that inventions were often "of no importance," because building up a new industry was fundamentally "a matter of building up an immense expertise and 'know-how.' "[83] Public funds ought to go to building a Fischer–Tropsch process chemical plant, one member argued in agreement, the purpose being "not necessarily for processes which are altogether new, but in order to gain the 'know-how' of treating coal in that way."[84]

By the late 1950s, the debate on developing and transferring technology (and thus promoting British exports) no longer focused just on patents and trademarks. Several parliamentary debates by the mid-1940s remarked on the growing importance of "that American term, 'know-how.' " In discussing the economic situation of 1947, Viscount Swinton felt the need to define this "American expression" as "invaluable intangible assets born of long and varied experience . . . [the] great aggregate . . . of thousands of individuals, individual enterprise, knowledge and, I would almost say, instinct." It was difficult to define precisely, but one would know it "by the smell."[85] By 1956, Sir Lionel Heald urged revision of a Restrictive Trade Practices bill to account for a practice "common nowadays, for there to be attached to . . . patent agreements an agreement for the exchange of know-how." Heald, quickly joined by other members, hoped to allow British industry both to learn from abroad and to package British "know-how" as an export good in itself.[86] As we will see, the timing of this rise of know-how in the United Kingdom fits precisely with the American experience, though the very fact of its discussion in policy debates and explicit inclusion in trade bills marks more government-level policy attention than know-how would receive in the United States for decades to come.

The Exploitation of German Science

This interest of members of Parliament in incorporating conceptions of "know-how" into patent and trademark law both mirrored and drew directly from the concerns of the BIOS administrators and CCG/BE officials. In the CCG/BE's view, the United Kingdom's future had "never before . . . been affected so vitally by export considerations," and the urgency of this problem undermined arguments for patents that depended on long-term incentives for research and development.[87]

The Board of Trade represents the main structural link between the British Zone of Germany, BIOS personnel and industrial investigators hoping to aid British exports and military technology, diplomats focusing on building ties with America, and members of the Houses of Parliament negotiating change to domestic patent law. Sir Stafford Cripps, president of the board from 1945 to 1947, was himself formerly one of Britain's most prominent patent lawyers. He answered parliamentary inquiries about the state of German patents and urged amendments to the patent law to indemnify companies utilizing BIOS information. Further, the worlds of industry and Parliament had considerable overlap. The honorable member for Heywood and Radcliffe, Mr. James Wootton-Davies, as just one example, drew upon his experience as an industrial chemist when arguing that a new patent law would be necessary for Britain to become a "scientific nation" and to aid inventors.[88] Other members drew upon their experience as patent lawyers, industrialists, and university professors in these discussions. Darwin Panel meetings generally had more members present from the Board of Trade than any other agency.

Until September 1947, the British Foreign Office left most considerations of German patents and trademarks, both within and outside Germany, to the purview of the Board of Trade, "in view of the extreme technicality of the subject."[89] This appeal to technical expertise was persuasive and consciously used, but by the late 1940s, the "urgency of expanding German exports" led to the British Foreign Office and other agencies insisting on involvement in controlling German patents.

In chapter 5, I go into more detail about developments in the German economy through around 1950, but much of the planning regarding German science and industry revolved around ensuring how to account for the anxieties mentioned earlier: British industrial decline, German cartels engaging in economic warfare, and missed opportunities if the state did not properly sponsor academic and industrial research.

Conclusion: Getting "A Bite Out of the Apple"

Sir Stafford Cripps, recapping the importance of BIOS as field investigations, neatly summarized the British viewpoint:

> All teams of industrialists will be out of Germany by June 30th (1947). For obvious reasons this date does not automatically end the transfer of "know how" from that country, it merely marks the end of the first stage i.e. active research in the field. The process of digesting the data procured must go on. . . . Only [then] can the final step, the integration of that knowledge in our manufacturing techniques—the ultimate object of the whole operation—be commenced.
>
> Probably the most important single aid to this process of integration is the employment of the key German scientists and technicians who were responsible for the particular advances in the industrial concerned.[90]

The United Kingdom likely had much more to gain from studying German industry than did the United States. British planners were also more unified in their planning and execution, perhaps as part of what historian Volker Berghahn has described as a "greater homogeneity of outlook [in Whitehall and Westminster] than in the executive and legislative branches of government in Washington."[91] With a smaller industrial base than either the United States or Germany and a lower cost in time and logistics to reach German facilities, we might expect some of the tremendous value from this "intellectual plunder" that popular histories describe.

Instead, the defining feature of the British experience was a struggle with the inherent difficulty of moving technology across geographic distance and industrial cultures, epitomized by the problem of getting "the 'know-how.'" British investigators pored over records, blueprints, and patents in Germany; interrogated scientists and technical personnel; and confiscated prototypes, and still the records are full of energetic complaints that they were not getting the technology they needed. Eventually, BIOS and similar groups resorted to know-how-based methods: hiring German scientists and technicians (despite sometimes fierce protests from local residents and unions) and embedding their own engineers in German facilities for weeks and months.

Once "taken," documents had to be processed, and once processed, they had to be sent to interested and needful parties. Yet German technology was not as broadly helpful as planners hoped. "The number of documents in any batch which are of real value to industry is very small," the head of one processing unit reported to the Board of Trade in February 1946, "possibly not higher than 5%. This fact cannot be determined from the title of the documents but only from expert evaluation."[92] A member of Parliament, while discussing patent reforms in 1953, commented that "it is now generally agreed that the results [of investigations in Germany] were disappointing, and that although the reports of the teams may have infringed copyrights, they added little to our industrial knowledge."[93] Perhaps there was, indeed, great value in the "negative information" that "in very many fields investigation has disclosed that our own technicians have little to learn from the Germans," but it is immensely more difficult to estimate if that was the case.[94]

The question of how best a state might aid industry in *making use of* new technology is one at the center of both BIOS-related schemes and the patent reforms that were under way in the United Kingdom in the 1940s and early 1950s. Both American and British policymakers were initially optimistic about their ability to overcome this gap between "investigating" and "using" (or, put another way, the gap between being told something and understanding it). They assumed that the "best" technology would simply win out and be a major boost to domestic industry. Thus, by publishing the results of German research, entire industries could benefit fairly and evenly. Faced with the challenge of putting this into action, however, British policy changed course to aiding individual firms through promoting longer on-site visits and establishing legal frameworks to license German technology. Helping individual firms might have seemed unfair, or at least open to corruption, but as one official argued, trying to help everyone would ultimately help no one.[95]

This ideological commitment to equal, worldwide distribution of the fruits of German research via published reports was not just an internal commitment—it was the basis of the reciprocity agreement between American and British investigatory programs. A loss of faith in the ability to help entire industries was just as much a loss of faith in learning from the American investigations and providing full value in kind. Later Soviet accusations of the United States and United Kingdom retaining the sole value from their investigations were certainly self-serving and somewhat unfair, but they were not particularly untrue. To the considerable frustration of British policymakers, German technology proved more useful for diplomacy than they had anticipated. A "bite out of the apple" was worth more than a "whiff all around," but it was considerably more difficult to make friends by sharing.

3

French Planning for German Science

Student Spies and Exploitation in Place

Emerging from the Second World War, the Provisional Government of France (which would become the Fourth Republic in 1946) faced enormous challenges: establishing a new constitution, rebuilding its economy, regaining international influence, and overcoming the shame of Vichy collaboration. Each of these challenges was made more complicated by internal politics. Communists, socialists, and the right-wing Popular Republican Movement vied for power in the theoretically unified government. Communist influence appeased the Soviets but made the Western Allies increasingly nervous, while the opposite was true about Charles de Gaulle's eventually successful demands for control of a powerful executive branch.

If communists, Gaullists, and foreign observers alike could agree on anything, it was that science and technology would be key in rebuilding France. Historian Gabrielle Hecht has shown that efforts to promote international leadership in nuclear energy were in part about forging a new identity for France as a peaceful, high-technology country.[1] The Nazi occupation had destroyed French science as much as any other institution, though. France's laboratories, universities, and factories had suffered serious damage. Many scientists and skilled technical workers had died on the battlefield, fled before the war, or in some cases even ended up in Nazi concentration camps. Some remnants of French science survived because Germany had hoped to use scientific ties as a symbol of allegiance with the Vichy regime, but little remained when the provisional government considered how science would serve and receive support from the new French state.

Reparations from Germany, including intellectual reparations in science and technology, were one potential solution to some of these challenges. Missions sprang up in the French bureaucracy to exploit German science and technology, especially in the French occupation zone. This, planners hoped, was a way to leap back to the forefront of science, to achieve, in the words of one French scientist, "the preeminence to which we pretend."[2] One estimate from the organization in charge of rebuilding French science put France some ten to fifteen years behind German technology and described "Le Probleme Francais" as one of building up a capable technical workforce.[3] To this extent, the French efforts fit neatly into the story of the unprecedented, international attempt at mass technology transfer enacted by the victorious power.

While the United Kingdom and the United States worked together closely on their scientific exploitation efforts, France remained largely on the outside. Only after extended debate and agonizing were American and British officials willing to conceded that there was more to be gained than lost by (partially) working with the French. To this end, a newly created FIAT (France) joined FIAT (US) and FIAT (BR) as a liaison agency. These FIAT branches authorized and coordinated the movement of teams of investigators between zones. A program run by the French military to study and reproduce V-2 missiles fairly closely mirrored Operation Paperclip. To France's allies, FIAT (France) (and related programs) seemed to be identical to theirs.

These outward similarities disguised essential differences between the US approach and that of the French Fourth Republic, especially regarding efforts to take German industrial technology and academic science. At its core, these differences emerged from what chief policymakers in each nation understood to be the possibilities and limits of technology transfer, combined with the distorted perspectives created by early Cold War diplomacy. If US efforts relied on an overly optimistic view that science and technology could be packed up in Germany like a microscope or book and dropped off in American industrial and military labs without complication, and the United Kingdom learned otherwise the hard way, France operated from an almost opposite set of assumptions. As key French officials saw it, there was no point in even attempting the systematic recruitment of German personnel or seizure of German labs or factories, because minds and even labs taken from their contexts were thereby rendered "practically sterile."[4] The only value to be had from them was in exploiting German minds right where they were—in the French occupation zone.

Working from this set of assumptions, FIAT (France) and its organizers continued to facilitate American and British investigations, but French efforts mostly shifted to emphasize *controlling* German science and rebuilding it to better serve French interests and follow French leadership. This meant rebuilding or founding German universities and labs in the French occupation zone as collaborative Franco-German institutions. In these institutes, French students would train with German scientists, both gaining skills and serving as de facto scientific spies. If hiring Germans was politically unpalatable, this method would at least more slowly help alleviate France's dire need for scientists by training *French* students with *German* skills. This was not an absolute doctrine—there are certainly important cases of the recruitment of German scientific and technical personnel by the French government. Nor was it France's first choice of policies. The ongoing diplomatic tension between France and its Anglo-American allies deeply shaped France's postwar science and technology policy by limiting what policymakers perceived as possible. Still, it seemed to be the best remaining option. This episode is a key example of how international relations as well as science and technology policy shaped one another in this era of highly politicized science. The French policy of scientific exploitation in place might well have both failed to achieve its immediate aims of control and been a great advantage to building Franco-German economic cooperation in the longer term.

Anglo-American Projections of French Intentions

Throughout the war, even though Roosevelt thought that (nominally) including Charles de Gaulle in the war was symbolically important, the Free France Forces remained firmly outside the confidence of either the Anglo-American team or the Soviet Union. This distrust continued throughout the provisional government and into the Fourth Republic. Even after the liberation of France, it was still not considered a true "power" in the same sense as the United States, United Kingdom, and Soviet Union. The Yalta and Potsdam conferences in 1945, in which the leaders of the Allied nations came together to plan out postwar Europe, consisted of discussions among the Big Three—not the Big Four.[5]

The joint Anglo-American Combined Intelligence Objectives Subcommittee (CIOS) was responsible for planning out what scientific/technical targets the T-Force units should secure for intelligence purposes. As CIOS learned of potential targets, it added the most promising and important leads to its secret "Black List," while less pressing or security-related targets went on the "Grey" or "White" lists. According to a CIOS memo written in 1945: "During the pre-invasion days and the Battle of France, various French agencies and individual officers rendered great assistance to Allied agencies in the collection of German technical and economic intelligence in France."[6] During the quick advances of 1945, the expulsion of Germany from France, and the formation of the French Provisional Government, however, a large number of Black List targets slipped by the T-Force troops. Now under the sovereign authority of this provisional government, investigative teams needed the cooperation of French authorities in order to reach and legally investigate them.[7] G-2 was placed in a bind between a desire to maintain secrecy from the French and the need for French cooperation.

On January 24, 1945, a G-2 officer in the Supreme Headquarters, Allied Expeditionary Force (SHAEF) Mission to France sent an official request back to headquarters in which it was "tentatively proposed" to send names of targets to the French government, including a copy of the Black List.[8] Approval—of sorts—came through. Brigadier General J. J. Davis of the Adjutant General's Corps of the US Army sent authorization with a strict warning: while the CIOS's Geographic Black List could be shared, the CIOS's "technical item" Black List "will in no circumstances be issued and care should be taken to ensure that the French do not become aware of its existence."[9] In other words, the United States was begrudgingly willing to share *where* they wanted to investigate within France, but what they hoped to find (and that they even had specific targets in mind) remained strictly secret from their nominal ally.

The French division of SHAEF sent an invitation to General Alphonse Juin, chief of the General Staff for National Defense in the French army, who was eager for the opportunity—so long as it was on French terms. Juin freely admitted that such investigations had "value not open to question" and were "greatly to be desired" but insisted that special instruction be given to avoid the impression that the United States and United Kingdom were investigating French industry. "Such an investigation into the workings of French concerns might run the risk of causing a certain uneasiness among the French industrialists whose plants were investigated, as well as among the public generally. Those who were insufficiently informed as to the real purposes of the inquiries might quite honestly interpret them as being a technical investigation of French industry."[10]

This was a reasonable concern. In general, the Nazi European economy (the "New Order") ended up giving French industry considerable leeway, so long as it produced what the German military needed. IG Farben, the powerful chemical conglomerate, was bitter over the French chemical and pharmaceutical industries having taken German patents and methods after the

First World War—in part due to France's notoriously weak patent system—and insisted on substantial indirect control over French industry.[11] French industrialists were able to resist German technicians being embedded in French industry to any significant extent, or new technologies otherwise being introduced, so there would be little Nazi technology for Allied investigators after the war—but there is no reason to believe that the United States, United Kingdom, or Soviet Union knew that (or necessarily cared).

Whether these remarks were a veiled warning against overstepping their bounds or a sincere effort to avoid misunderstandings, Juin was sold on the idea of cooperation for the exploitation of German industry. In the same letter, he introduced the newly created Comité de coordination scientifique de la defense nationale (CCSDN). This group would be an analogue of the joint US–UK FIAT and indeed soon became known as FIAT (France), even using this nickname on its own stationary. Juin finished his response by suggesting that this collaboration extend beyond the border of France and into Germany itself, in a truly cooperative and integrated intelligence effort.

This was a step further than the Anglo-American planners were willing to go, at least without significant hedging and drawn-out discussion. The members of SHAEF suggested sharing the Black List with France in January 1945 yet only received permission for the partial list in March. The first response to Juin's suggestion of closer collaboration came in May, when the Combined Chiefs of Staff in Washington pushed the matter off still further, saying that they "consider[ed] it inexpedient to take action in Washington relating to details of internal administration" and would refer the matter back to SHAEF (who had referred the matter to the chiefs in the first place months earlier).[12] In mid-June, US Adjutant General Davis agreed in principle to a reciprocal arrangement in which express authority would be required from an occupation zone's administration before investigators could be sent or files copied, then again insisted to SHAEF (France) that the CIOS technical item Black List remain strictly secret from the French.[13]

French Army General Darius Paul Bloch-Dassault, president of the CCSDN, was among several French officials who registered complaints about this continued formal exclusion from the Anglo-American technical intelligence missions. He insisted that they be represented on the CIOS committee, criticized the "attitude" of SHAEF, and generally lobbied to "participate fully with the British and US Forces in the collection of technical and economic intelligence."[14] Delays continued. In June 1945, the US and UK intelligence units insisted that no agreement could be completed prior to the upcoming dissolution of SHAEF—fully eight months after the G-2 first suggested cooperating with France in technical intelligence. Even then, the matter was "still under consideration by higher authority" than G-2 itself.[15]

Late August finally saw some progress in integrating the French, if only in deciding what limits to place on cooperation. FIAT (BR)—now a separate, national entity with the dissolution of SHAEF—decided that "no release of information from the FIAT Secret records . . . be made to the French. Information from the non-Secret records," however, "may be given."[16] Visits to the French zone could be arranged through the French liaison officer to FIAT (BR). Authorities from the United States clearly agreed with this assessment of how much (or little) to trust the French, as the deputy military governor for Germany to the Joint Chiefs of Staff, in a November 1945 memo, requested that the long-standing ban on sharing secret files with the French be extended to Soviet requests.[17] In the early occupation phase, it seems, the French were even less trusted (or at least less important to placate) than the Soviets.

The ongoing debates over official policy on how far to trust France to join the scientific exploitation efforts did not prevent unofficial cooperation throughout much of this period. Hundreds of investigative teams from dozens of organizations and missions streamed into occupied Germany: military intelligence branches, government-sponsored industrial investigators and recruiters, the Strategic Bombing Survey, Technical Oil Mission, Alsos Mission, UK Board of Trade–sponsored groups, and additional French agencies discussed later on. Informal arrangements allowed teams to receive passes to investigate targets within each zone, with liaisons from the national FIAT services coordinating logistics. Even before General Juin's initial letter requesting formal inclusion, Juin knew of several occasions on which "such a collaboration . . . has already been put into effect."[18] Throughout 1945 and 1946, all involved emphasized the importance of "informal contact on technical points" and a "desire to see personal contacts maintained" and "the present cordial and effective relationship . . . strengthened."[19]

The lack of formalized, sincere coordination and cooperation left room for clandestine action. The French military and secret service was no more afraid to make aggressive bids for technical personnel than were the other nations investigating Germany. In May 1945, the same month as the creation of the CCSDN (aka FIAT [France]), de Gaulle—prompted by an optimistic note from General Juin—issued a top-secret order authorizing the "transfer" of German scientists and technicians "of great value" to France for interrogation and to "engage them to stay at our disposition."[20] The details of how to determine this "great value" were not discussed, nor were the limits of how to "engage them to stay."

The American and British agencies involved in exploiting German science and technology certainly felt aggrieved by anecdotes of French teams sneaking in and "stealing" promising personnel, offering them more lucrative contracts or freedoms than they had been offered by the United States or United Kingdom. In June 1945, Brigadier General Eugene Harrison, assistant chief of staff of G-2, sent a file to his superiors detailing French "violations" to date.[21] A German radio engineer reported that the secretary of the French Committee for Industrial Production had told him in December 1944 "that he was not to reveal to the Americans or British any details of German radio and radar technique, manufacturing methods, or research." Dr. Kurt Wilde, the director of the German firm Askania, claimed a similar warning from the French navy, stating that he had been threatened with being added to the Allied list of war criminals if he were to share his knowledge with the Americans or the British. Several other CIOS-sponsored teams reported similar instruction from the French Securité militaire, Navy, Department of Industrial Production, and the Arsenal. At this point, Harrison added that "difficulties in the French area have been very few and have been largely limited to the field of electronics," but nonetheless this information was significant enough to merit a note handwritten on this report by its recipients: "A special file on French activities of this nature should be kept."

Whether for this report or not, incidents accumulated in American and British records of French "misdeeds," perceived or real. When the French detained the chairman of the board of Messerschmitt, a prominent German aircraft company, along with cartons of the company's records, an American offer reported suspicions "that if this incident had not been witnessed by an Allied officer, the existence of this material would not have been made known to the Allies."[22] Separate reports from February, May, and July 1946 tell of several incidents of scientists hired by French agents despite their being in the custody of British or American FIAT. In one case, V-weapons researcher Hermann Ehrenspeck was jailed for threatening to go to the French zone without permission rather than surrender his papers to FIAT for inspection.[23] The American State-War-Navy Coordinating Committee—tasked by the Joint Chiefs of Staff with developing "long-range Government policies and procedures" regarding the exploitation of German scientists—drafted a report in March 1946 warning that if efforts were not quickly escalated, the Russians were "already proceeding with an aggressive policy of long-range exploitation" and the French were "offering lucrative contracts to selected specialists."[24] Similarly, the US Army's Air Intelligence Division issued a report in April 1946 listing ten "typical" cases of exploitation by the "French and Russian agents," with which the American zone was "literally crawling."[25] A State Department official reported in June 1946 that they had "caught the French red-handed again stealing scientists out of our zone."[26]

It is important not to extrapolate too far based on these American and British reports. To be sure, there is little reason to doubt that the French military and security services recruited specific German scientists, at least to a limited extent—this is the focus of the next section. However, these anecdotes were also very useful for FIAT (US), BIOS, and other agencies involved in taking German technical secrets. These reports were used to promote the construction of detention camps for German scientists and technicians (known as Dustbin and Ashcan) and to argue that increased funding and attention be given to existing scientific exploitation efforts. Americans and the British argued, basically, that they had better amplify their own efforts, because everyone else was already doing it, and they would not want to miss out in this zero-sum game of scientific and technical war booty.

These other Allied nations might have been fooling themselves as much as any superior. Their preexisting distrust of the French Provisional Government combined with these reports of nefarious French espionage to reinforce an impression that the French were playing the same game as they were, just in a more underhanded way. That was largely a misunderstanding, and one that both created and reflected diplomatic tension among the Western Allies.

French Policy and Perceptions of Technological Embeddedness

Certain policymakers and powerful institutions within France—especially the military—saw tremendous potential gains to be had by exploiting German science and technology. They operated a systematic effort from 1945 through at least about 1950 to target scientists and skilled technicians and bring them to France. Groups involved included representatives of the three military services, in the form of the Bureau Scientifique, Guerre; the Mission Intelligence Scientifique et Technique, Air; and the Section de Liaison/Recherche, Marine. The results of this effort were significant, particularly in the emerging field of military aerospace technology (and, even more specifically, V-2-based ballistic missiles).

Enthusiasm for German military technology was every bit as strong among French security planners as it was among other nations' militaries. On May

16, 1945, the chief of staff of the Defense nationale send a top-secret message to General de Gaulle, reporting that "the results obtained, especially in the domain of secret weapons, has very much impressed those who have examined them," adding that certain personalities had already been brought consensually to Paris for interrogation.[27] De Gaulle quickly agreed that it was a fine way to proceed. By November 1946, the minister of armaments reported that it had imported about eight hundred technical personnel.[28]

Most of these specialists likely worked with the Groupe opérationnel des projectiles autopropulses, the subbranch of the Ministry of Armaments tasked with learning how to build German V-2 rockets. This group, led by Professor Henri Moreau, began their efforts even before the occupation of Germany by collecting parts of V-2s that had been used on and around Paris. As the opportunities arose, they moved on to investigating the plants, both on French soil and in the French occupation zone, in which the rockets were built and tested. Finally, they progressed to recruiting German personnel who had worked on the V-2 rockets.[29] For Moreau and "certain military authorities," as Jacques Villain writes in a history of the V-2's legacy in France, "it was clear that, before undertaking to produce the missile, France should first assimilate the German know-how. . . . [It] was also clear that it was necessary to obtain the services of the German specialists who had designed this equipment."[30] Villain does not provide a final number for the personnel recruited for this expansive and decades-long effort to harness the information gleaned from the V-2 and its designers, but at least 123 technicians were hired just from the Peenemünde alumni (the von Braun group mostly taken to America during Operation Paperclip). The V-2 rocket had a major psychological impact, despite its relatively modest influence as a functional weapon during the war, and France was as eager as any of the victorious powers to expand on its apparent potential.

Certainly hundreds and perhaps thousands of German scientists were brought to France in these years. In some ways, this is relatively small scale compared to the American or Soviet V-2 programs, but the French were also not in an economic position to put the results of any program into mass production regardless. Not only did most of France's armaments industry lay in ruins but 62 percent of remaining armaments factories were converted to produce desperately needed tractors, machine tools, and other civilian industrial tools.[31] In 1947, a new five-year plan envisaged rebuilding the armaments industry, including aeronautics, but before this remained an unfunded ambition, rather than a reality, until at least 1950. The French armaments industry did slowly rebuild in the 1950s onward, but there was nothing akin to the American or Russian ballistic missile crash programs in the immediate post-war years, when they could directly cash in on hired German expertise.

The V-2 was prominent but far from the only military technology of interest. As various branches of the military identified technologies they wished to be investigated, they coordinated with the civilian agencies discussed later in this chapter, the occupation government, and sometimes the covert action wing of the security service in order to import both prototypes and person-nel. When the navy wished to learn more about a lab that studied using mag-netic fields for undersea mapping and detecting mines, they worked with these groups (and also the air force) on moving that lab, its head, and six to seven key researchers to Marseille.[32]

While the French military was eager to acquire German technology by bringing in the technical specialists and setting up camps in French territory, very different thinking prevailed among the groups tasked with taking Ger-man civilian industrial technology. Differences between how these agencies and their American and British counterparts conceived of technology and its role in society, in turn, exacerbated diplomatic tension among these Western Allies at a crucial period in the emergence of the Cold War.

Like in the United States and the United Kingdom, many different French agencies had some role in the exploitation of German science and technol-ogy. In fact, the bureaucratic lines could become very difficult to untangle even at the time. A FIAT (France) report on July 4, 1946, complains that no fewer than three separate agencies possessed, in theory, a monopoly on com-merce between France and occupied Germany, which somewhat frustrated several others with mandates to import goods and hire personnel from these areas.[33] Among the prominent agencies with at least a potential mandate to exploit German science and technology for civilian industrial purposes, the most active were the Centre national de la recherche scientifique's (CNRS) mission to Germany; the branches of the Production industrielle, Économie nationale, and Education nationale tasked with the occupation zone; and the CCSDN. The secret intelligence agency, the Direction générale des études et recherches, which later became the Service de documentation extérieure et de contre-espionnage, also played an important if limited role: providing pass-ports for desirable German scientists; making covert contact with key Ger-mans in American, British, or Soviet custody; and otherwise facilitating the other groups' missions when needed.[34]

These were not fully independent entities, nor were their relationships stable. The CNRS mission was at one point in 1945 attached to FIAT (France), but the director of the Education nationale thought that it should be under his purview, leading to an administrative battle that remained a seed of resentment even after a nominal compromise involving a shared deputy position.[35] Several additional restructurings took place in 1946 alone. Finally, FIAT (France) was absorbed in April 1947 into the relatively new general intelligence agency, the Service de documentation extérieure et de contre-espionnage.

Of these, two are particularly important: the CNRS mission sent to the French occupation zone and the CCSDN. The CNRS itself was founded in October 1939, right around the start of the war, in a bid to centralize both basic and applied scientific research. Its original mission was to organize reparations seizures of scientific equipment from Germany and to consult for the occupation authorities on how to control German science. This mission sprawled over time. By 1947, it also facilitated orders of scientific equipment from German manufacturers for French laboratories, organized exchanges of scientific personnel between France and Germany, ran the French side of the *FIAT Reviews of German Science*, and maintained contact with German scientists (including exchanging periodicals) for both diplomatic and intelligence purposes.[36]

The CCSDN was originally founded on April 20, 1945, to help coordinate the military departments' research. It was to receive reports from each branch, assign topics of mutual interest to one group or another, and otherwise reduce duplication and improve communication. Early on, however, it was also tasked with researching enemy science, technology, and industry, both in terms of providing assessments to military leaders and actively acquiring those technologies for French military and industrial benefit. When the opportunity arose to coordinate with the American and British exploitation programs in occupied Germany, it was a natural fit. It thus became known as FIAT (France), though its mission was substantially broader than FIAT (US) or FIAT (BR). The membership of FIAT (France) included representatives of the War, Air, and Navy commands; the research branches of each of these commands; the Direction générale des études et recherches; and a scientific counselor from CNRS. The vice president of the CCSDN was always the director of the CNRS. In this director's seat sat Frédéric Joliot-Curie, son-in-law to Marie Curie and himself a future joint Nobel Prize awardee for radiation research.

In CNRS and FIAT (France) board meetings, the basic questions debated about had been familiar to their allies: Should German science be destroyed?

Should it be captured and brought home? Should they focus on crushing the Germans so they never rose again, or rehabilitating them so they never *wanted* to wage war again? How could French industry and science gain the most possible from the riches of German technology? However, among the leaders of these nonmilitary (or, at least, mixed civilian/military) agencies, a very different line of thinking about the limits of technology transfer prevailed.

French officials ultimately decided that moving scientists, engineers, or even entire factories or laboratories from one place to another would strip of them of their value to such an extent that it was not worth the cost. Technology, as part of society, was no shiny bauble to be picked up and carried off. A report written by the Service for German Affairs of the Économie nationale titled *The Problem of German Scientific and Technological Research* (dated March 10, 1946) is just one clear example of this viewpoint.[37] The looting and destruction of German capacity for scientific research was tempting, the researchers argued, but in practice it was unrealistic to hope that this would benefit France. Destroying German laboratories would be a setback, but Germany would not be occupied forever, nor would France be in a position to impose its will on the other occupying powers even in the interim. The Bureau Scientifique de la Armée disagreed and presented a thorough plan to repress German science, but that seemed both impolitic and impractical to most of the civilian planners.[38]

Meanwhile, grabbing German scientists and bringing them into France was equally counterproductive. The "milieu" around a scientist, including personal contacts, is so essential that "a top scientist or technician displaced in another country . . . is practically rendered sterile."[39] Transferring such a person, "by cutting all his connections, seriously diminishes German science . . . [and] he does not regain his dynamism until after he has established new connections with the French scientific community. . . . It is therefore preferable to let him reside in Germany and to control his activity in a manner of use to the French state."[40] "In general, the transfer of German laboratories and centers of research would cause a considerable diminution and render laboratories of little value and difficult to integrate into the French economy."[41]

The Économie nationale was only one of several agencies arguing this point. During the following meeting of FIAT (France), Engineer General P. T. T. Janes, head of the Centre national d'études des télécommunications (National Center for Telecommunication Studies), stated: "The principles enunciated [by the Économie nationale] are excellent, in particular the national problem is perfectly posed."[42] Engineer General Gaston de Verbigier de Saint Paul, head of FIAT (France), echoed the Économie nationale's report almost verbatim: He "[felt] that the transfer of research establishments to France considerably diminishes what they can give us. Researchers work in <u>teams</u> and it is pointless to hope that some technicians taken from their establishments could continue their work in good condition."[43]

These were not simply an argument that transfers would be difficult or hiring Germans politically embarrassing, though those arguments were brought up as well. This is a conception of science and technology as profoundly embedded in society. With such a different conception of the possibilities of technology transfer, new methods of harnessing German science and technology would be needed for these agencies to fulfill their missions. They settled on a policy of controlling and building up German science in place but rebuilt in ways that would benefit French science and industry long term. First and foremost, this meant getting French trainees, or *stagiaires*, into German research facilities.

The Stagiaires Plan and Scientific Control from Within

If scientific prowess could (and should) be neither destroyed nor effectively transplanted, as the CNRS mission and the Économie nationale suggested, it would be necessary to turn German science to the benefit of France right where it was. Around May 1946, the CNRS outlined a plan discussed at length with these other agencies over the previous weeks: essentially, French stagiaires would be placed in German laboratories to complete their training.[44] Though a simple plan, its objectives were ambitious: "The problem of limiting research is, above all, a problem of control of training and control of what is studied."[45] By embedding French trainees in German labs, the proposal went, France would have a direct window into exactly what type of research those labs were pursuing.

This reflected persistent fears that German scientists were still secretly pursuing dangerous military weapons but hiding their violation of quadripartite controls by playing the occupying powers against one another. The fear of German military resurgence sometimes led to fairly far-fetched theories of German scientists splitting up research on the next superweapon into four individually innocuous pieces, pursuing each across zonal lines, and then hoping to put it together in the end.[46] Less drastically, many CNRS and FIAT (France) meetings had at least one representative who worried that "controls haven't been enough to prevent forbidden research."[47] The stagiaires would, they hoped, provide a crucial on-the-ground, long-term control too closely integrated with the research process to be fooled by clever report-writing.

More importantly, though, these stagiaires would acquire the know-how behind Germany's vaunted science and industry in the only way possible: through sustained, hands-on, in-person training and experience. Rather than importing German scientists to rebuild France, it would be German-trained French scientists—who therefore had the deep connection to French society and culture needed to be effective—who would lead the way.[48] Even those who believed that there would be benefit from investigations and report-writing akin to the American model conceded that France lacked the technicians and scientists necessary to absorb German technology effectively, meaning that training technical cadres was the top priority.[49]

This plan quickly gained support outside of the CNRS and the Économie nationale, but the types of objections raised to it underline French concern about the social nature of technology. In December 1946, M. Wurmser of the Commission for Foreign Affairs raised a concern that generated a lengthy debate: Who should these stagiaires be? Perhaps, Wurmser argued, they should only send abroad advanced scientists and technicians who were "already formed," as those too young might learn "more than just techniques" from their new German colleagues.[50] Additionally, those spending the start of their research careers abroad would not be able to "work profitably" upon returning to France. Finally, they might lack the expertise to perform their role as de facto intelligence agents by recognizing the signs of illicit German research. Conversely, sending established scientists would undermine the attempt to learn from the Germans and would deprive France of desperately needed expertise.

An additional concern always on the minds of French planners was how their actions would reflect on the prestige of the recently occupied, rebuilding nation, and this was an issue in regard to scientific reparations as well. The representative of the École nationale supérieure des industries chimiques even argued that it was better not to try to track German science than it would be to choose poorly in selecting administrators and stagiares, since being fooled would hurt French prestige.[51] M. Rapkine, who had previously served as a sort of wartime scientific ambassador to the United Kingdom and helped smuggle scientists out of Vichy France, recommended sending people "intermediate" in their training as a compromise. They would be able to oversee German science and would benefit at least somewhat, yet they would remain inherently French. More or less satisfied with the compromise, the committee agreed to pursue this plan under the auspices of the CNRS mission in the French occupation zone.

Results of the stagiaires plan were underwhelming from a control standpoint. This effort received its first reviews in a meeting of the High Commission for Research in Germany, held in March 1947 (a commission that included representatives from most of the groups operating in the field of exploitation of German science). The stagiaires, "whose official purpose [was] to learn German methods, but who had the aim of 'pumping' information about German labs," "could not take part in any activity of this genre," having been placed in a position wherein no one would confide in them and they learned little.[52] The trainees simply were not being included in any lab decisions that might be secret, since they were seen as agents of the control authority. By October 1947, seven months later, this scientific intelligence aspect of the plan was abandoned. In a meeting on October 13, General Jacques Humbert, speaking for this committee, "remarked that the training of French students and engineers in German laboratories didn't present any real value from the point of view of control," despite high hopes in the early planning.[53]

Results in other areas were far more promising, if harder to measure. There is reason to believe that these efforts built lasting bonds, both academic and industrial, between the two nations, even beyond what was recognized at the time. In a history of Franco-German industrial cooperation from 1945 through the 1960s, Jean-François Eck shows that even though the exploitation programs disappointed private industry at the time with their "mediocre results," the "interpenetration of interests" forged during this period set the stage for a surge in joint ventures from the late 1940s onward.[54] Growing Franco-German industrial and scientific ties by the 1950s—an incredible achievement given the animosity engendered by two wars and occupations—in turn played a key role in the development of the European Union. As historian Gérard Bossuat argues: "Franco-German industrial cooperation was justified by the construction of Europe. . . . Franco-German relations where at the heart of the success of the modernization of French armaments."[55] One of the labs initially considered for transfer deep into France but left in place to become the Franco-German Institute for Research at Saint-Louis is one example of this renewed cooperation in both civilian and military research.

This French planning around using scientists to build influence and gain intelligence also helps explain some contemporary French suspicion of others doing the same. In November 1947, the CNRS entered discussions with both British and American scientific societies to set up exchange programs.

On the UK side, this resulted in the establishment of the Advisory Council on Science Policy, but talks dragged on without much result. Attempts to build closer ties to the Swiss through science ran aground of problems finding someone to coordinate with—Swiss science was too decentralized and lacked a CNRS-like agency. Despite these outreach efforts of their own, though, French planners were worried when others reached out to them. When the Federation of American Scientists offered laboratory spots for European scientists and workers, this led to suspicions that it might be a ploy "intended to carry off Europe's scientists."[56]

Much has rightly been made of the importance of science in American "soft power" diplomacy in the postwar years, including in the reconstruction of Europe.[57] John Krige has compellingly demonstrated that American institutions (public and private) used science and technology as tools for knitting together an American-oriented, international (but anti-Soviet) network of scientific and technical elites in Western Europe from the 1940s onward while relying on the nominally "apolitical" nature of science as a shield against accusations of imperialism or meddling. This was particularly useful in recruiting the likes of Joliot-Curie, director of CNRS and part of FIAT (France), whose overt communist political leanings made him suspicious of American influence. However, it was not just the United States using science as soft power. While the United States had the resources and willpower to play this game most aggressively, France, too, ended up turning its exploitation programs and postwar science policy toward building up diplomatic goodwill and stronger industrial ties.

In the following chapters, I go into more detail about how these French collaborations with German academic science were, in part, a way for France to rethink its own domestic science policy. This world of academic science, in turn, was an important part of building the Western European community of nations that became the European Union. Because science was often considered apolitical, it was an ideal venue for collaboration, especially as particle accelerators, nuclear reactors, and space programs seemed too expensive for small and medium-sized nations to pursue on their own.[58]

Exact budgetary numbers spent on exploitation are difficult to find and would be even more difficult still to compare across these nations due to the challenges of trade and currency conversion. One of the few exceptions is in the CNRS mission to restock French laboratories with physical equipment either taken as reparations or bought from occupied Germany. Initial planning was for these seizures to be counted as "war booty," which did not even count against each nation's reparations tab. The CNRS took orders from labs throughout France for what they needed and began collecting from German labs and placing orders in factories in the French occupation zone. However, political ties to the Ministry of Finances soured in 1945, and the minister withdrew permission for the CNRS to import goods from the occupation zone as reparations or war booty. Instead, the CNRS could only bring goods into France at a firm price of 80 percent of the world market rate.[59] Labs and companies had placed orders for equipment expecting it to be free of charge and so complained when the CNRS attempted—with great difficulty and limited success—to collect this money.[60] In 1947, the acquisition of scientific equipment from Germany and Austria cost about 11 million FRF, which totals to roughly 800,000 USD in 2015 currency.[61] For a nation desperately attempting to rebuild in many different ways, this was a serious outlay.

Still, even with these purchases and reparations seizures, the overall French intellectual reparations programs were almost certainly far cheaper than the American approach of sponsoring industrial investigative teams, copying field reports and patents, shipping prototypes, transporting technical personnel and their families to the United States even before finding employment for them, and having a branch of the Department of Commerce dedicated to advertising and distributing this information.[62] These savings, for example, might well have made possible the missions sent by the CNRS and technical groups to the United Kingdom and the United States, which yielded great benefits to French science and industry.[63] Yves Rocard and Louis Rapkine, scientists of some renown in France, each spent time on missions to the United Kingdom, not only performing the duties of a modern scientific attaché (e.g., fulfilling orders for journal subscriptions and arranging housing for visiting scientists) but also investigating British industrial technologies and helping make commercial connections to French industry.[64]

There were real costs to the French approach as well, however, especially on the diplomatic front. Focusing on exploitation in place meant not spending time and resources on mirroring American methods, which meant confusion and suspicion. American and British investigators were allowed into the French occupation zone, but when these investigators requested copies of documents, the French authorities treated this as a low priority, upsetting these Allied nations. In the view of FIAT (BR), "the difficulty at the moment is not so much that the French are unwilling to let us or the Americans, or for that matter anyone else, have a look at [French zone technical information], but that they have not made any attempt to co-ordinate it in such a way that their own industrialists or the other interested powers can use it. So far as we and the Americans are concerned, the French collection seems to be the only substantial block of data that is likely to escape us."[65] American and British representatives debated through the late 1940s "whether they [the French] have valuable information which they wish to keep to themselves, or whether they think we can be lured yet further with what may prove to be mediocre bait."[66] The idea that French policymakers could simply not believe in the cost-effectiveness of splicing German technology into France's economy does not seem to have occurred to the other major powers.

Conclusion

The international scale and ambitions of the Allied experiments in scientific intelligence is itself remarkable. The American branch of FIAT "processed over 29,000 reports, confiscated 55 tons of documents, and made over 3,400 trips within Germany . . . through June 30, 1946."[67] The United Kingdom issued a press release that was reprinted in the *Daily Express*, claiming to have investigated twelve thousand targets and written twelve hundred reports (American reports, they suggested, "are more exhaustive but not so highly selective as the British") by October of that same year.[68] Statistics for the French side are more difficult to come by, but in August 1946, FIAT (US) recorded about one-third as many French investigators and teams as British ones touring the US zone (33 French teams compared to 101 British teams).[69]

In efforts to copy military technology, French generals appear to have agreed with their British peers who were "whole-heartedly behind the whole operation."[70] Considering the importance of novel, science-based technologies such as radar in fighting and winning the war and the psychological impact and apparent potential of less mature weapons such as V-2 rockets or jet airplanes, this makes sense. The diplomatic cooperation is worth noting in and of itself, given how deeply the American and British governments distrusted both the Gaullist and communist factions in the provisional government of liberated France.

Civilian technology and academic science are a very different story, however. To be sure, France still had much to gain from Germany in these areas, especially when it came to physical reparations (lab equipment, machine tools, reagents, etc.). German scientific and technical personnel offered expertise of real value to a French nation seeking a resurgence of military might and international influence. Despite these incentives, key French policymakers differed drastically from their British peers, who saw the military investigations as "very conveniently adaptable to deal with those longer term peacetime interests of science and industry," and the expansion into civil industrial technology as "a natural extension" of such efforts.[71] While France created agencies analogous to FIAT (US) and BIOS, most of its civilian exploitation agencies pursued very different goals by very different means. The chief reasons seems to be a fundamentally different understanding of the social embeddedness of science and technology, and therefore the possibilities and limitations of technology transfer.

Naturally, there were people who disagreed with the strategy of exploitation in place, including important civilian policymakers. The commissioner of the Republic of Strasbourg wrote to the minister of the Production Industrielle in November 1945 eager to accommodate German industry reinstalled into Strasbourg. He even supplied a list of industries he thought would fit in well: chemical products, aeronautics, electrical equipment, nail manufacturing, and automobile factories.[72] Their voices lost out or were excluded from the exploitation planning, but they certainly existed. At least part of what made the French different from the American case was the relatively lower level of influence that industrialists and lower-level policymakers had over these decisions, in addition to any kind of overall different take on technology and society.

With little analogous to the Anglo-American "special relationship," French policymakers had little reason to override skepticism of the other Allies' approaches to technology transfer. Given French policymakers' understanding of science, they had few choices when it came to keeping Germany from rebuilding its military might. Removing German scientists was wasteful—to transplant a scientist or even an entire laboratory, they believed, would be like hacking off the roots when transplanting a tree. German labs could be occupied for months or years but not permanently, so tearing down German institutions was a temporary solution at best. Instead, they would have to pursue a policy of control in place. This approach, they admitted, would not accomplish the ideal goals of entirely co-opting or destroying German scientific potential. Still, it was a practical path toward real benefits. Students would gain skills and spy on German research, and German researchers would remain employed in peaceful research in the French occupation zone instead of disappearing to Soviet or American labs.

Ultimately, despite the failure of using students as intelligence agents, French plans might well have done more for French industry long term than any of the other Allied powers. American and British struggles have already been discussed, and Soviet paranoia and aversion to crediting "bourgeoisie" German experts severely restricted their exploitation programs. The active promotion of Franco-German scientific institutes and business relationships, meanwhile, built important ties between otherwise deeply hostile nations. The development of the European Union is a complex history beyond the scope of this chapter but depended heavily on industrial cooperation (especially in coal, steel, and, to a lesser degree, chemicals).[73] For a nation that so prized rebuilding its reputation as a legitimate world power, a role at the heart of Western Europe was likely worth far more than boxes of reports and miles of microfilm.

4

Soviet Reparations and the Seizure of German Science and Technology

For many German scientists in the immediate postwar period, the Soviet Union was a savior, extending a rare opportunity in a war-shattered landscape to provide for their families by applying their hard-won technical skills to interesting projects. For many German scientists, the Soviet Union was a terror, forcing them to abandon their homeland and move deep into Soviet lands with little hope of escape until Soviet scientists had fully absorbed their know-how.

"Intellectual reparations" is a useful phrase to describe the exploitation of German science, but for the Western Allies, the connection to reparations, per se, is somewhat indirect. Teams aiming to take German science and technology rarely registered their actions with the quadripartite reparations agency (the Inter-Allied Reparations Agency), and they operated mostly outside of its organizational structure. This was useful for the United States diplomatically, since it was able to posture as having been magnanimous and not largely uninterested in reparations, in contrast to the Soviets taking billions of dollars in industrial plants. This contrast—between public statements about the United States not taking reparations and the reality of intellectual reparations— is at the heart of John Gimbel's *Science, Technology, and Reparations*, the best source exploring the US side of this topic, and indeed at the core of most scholarly discussion about the topic.

In Soviet policy, though, German industrial technology, forced labor (technical or manual, skilled or unskilled), military equipment, scientific expertise, food, and other basic goods were all closely tied together. Soviet needs were tremendous, and rebuilding Eastern Germany was a low priority relative to reconstruction of the Soviet industrial and agricultural base. The result was a willingness to enact the violence and overt control over German industrial and military technology—and its inventors—that the Western Allies rarely needed or chose to employ. Science and technology were fundamental to Soviet goals in East Germany, even if it was often material versions of that technology (e.g., factory equipment or prototypes) rather than the immaterial (e.g., patents, data, blueprints). "Intellectual reparations" were truly part and parcel of a larger set of reparations policies in the Soviet zone.

Comparing the Soviet case to the Western Allies has some real problems. As a top-down control economy, technical expertise played a fundamentally different role in the Soviet economy and in its society more broadly. The Soviet patent system was largely unlike that in the capitalist world, offering direct payments as awards for important inventions rather than granting a period of personal monopoly over economic use of an idea. Science was deeply political the world over, but that, too, had a very different inflection in Stalin's Soviet Union. On the one hand, in Soviet ideology, communism was itself a science, and large-scale engineering projects were central in projecting Soviet power.[1] On the other hand, especially in the Stalinist era, scientists (especially those trained in prewar Germany) were vulnerable to getting wrapped up in paranoid internal politics and accusations of being ideologically impure.

Many of the differences between Soviet and the Western Allies' exploitation of German science and technology are issues of magnitude rather than kind, despite these overall structural differences. Soviet exploitation of German science and technology included more overt violence and coercion than its allies generally employed, and each incidence was multiplied in Western media and rumor mills among the fraternity of German scientists. American, British, and French forces also jailed German scientists at times. The Soviet Union dismantled enormous amounts of industrial facilities and research centers in their occupation zone and shipped them east, often creating tremendous waste. Even there, though, we can see parallels in the ineffective overreaching of FIAT and BIOS microfilming teams. It is important not to understate the suffering that Soviet rule imposed on citizens of the future East Germany, but the parallels across even these vastly different economic and political systems can be telling.

In this chapter, I explore the Soviet attempts at exploiting German science, both in the immediate aftermath of the war and how they—like the British and French—built infrastructure to benefit from German science long after the occupation. This chapter builds primarily from a synthesis of existing histories of Soviet exploitation of German science, rather than new archival research, and as a result will be of less interest to specialists in Soviet history.[2] For readers less familiar with this history, the Soviet case provides an important perspective on what the other powers were attempting, and how they went about it.

Soviet Reparations Policy in Germany

The Soviet Union paid a staggering price for victory in the Second World War. Casualty estimates vary, but a common figure is twenty-five to twenty-six million Soviet citizens, or 13 percent of the population in 1940.[3] In retreating from the Nazis, the Soviet army burned everything along their path to deny their enemies any resources. The Nazis razed buildings and enslaved people who survived. Finally, the Soviet counterattack added yet another layer of destruction. The initial Nazi assault through Poland and into the Soviet Union set a horrific standard for conduct, with widespread war crimes including rape, murdering civilians and prisoners of war, and indiscriminate looting. In Nazi ideology, this was land that needed "excess population" removed for future settlement by Germans, so there were few attempts by Nazi officers to rein in these atrocities.[4] This, in turn, set the stage for horrific retaliation, as Soviet soldiers fought their way through the gruesome remains of these crimes and into the lands of those who had committed them.

In the face of this destruction, even extreme Soviet demands for reparations seemed reasonable to Allied leaders. As Churchill wrote to Roosevelt in November 1944, "Uncle Joe [Stalin] certainly contemplates demanding two or three million Nazi youths, Gestapo men, etc., doing prolonged reparation work and it is hard to say he is wrong."[5] Science and technology (especially military technology) were among the most important targets of Soviet seizure. Understanding the efforts to gather them requires discussing the context of the Soviet occupation of Germany as well as the widespread industrial dismantling, destruction, seizure, and finally exploitation in place that characterized Soviet rule.

Soviet leadership and troops on the ground shared a desire to make Germany pay for the war. Planning leading into the Potsdam conference settled on a rough goal of 10 billion USD in reparations, combined with access to the coalfields of the Ruhr region, a chance to learn from German science and technology, and an end to any possible revived German military aggression.

Later, political goals came to the forefront: propping up an East German communist party that would follow Soviet instruction and framing the occupation as a "liberation" from Nazi rule.

There was little in the way of unified planning on how to achieve these goals, however, and Soviet occupation "policy" was riven by internal feuding and political rivalries. Even basic lines of command within the Red Army lost much of their power, as soldiers immersed in propaganda about the inhuman Germans—and who had just marched through the lands recently occupied by the Nazis—first saw the relative wealth of German lands. Soviet soldiers raped and looted in appalling numbers, and they and their low-level officers ignored efforts from generals to stop them.[6] Marshal Georgy Zhukov, the man in charge of the Red Army in general and the Soviet occupation of Germany in its first years, attempted to institute an extremely strict system to punish this disregard for authority. Zhukov found himself overridden by Stalin, who seems to have preferred that no centralized, organized source of authority exist other than himself, especially that far from Moscow.

These internal rivalries (e.g., the one between Stalin and Zhukov) led to inconsistency at any given moment and to major swings in policy over time. Like in the United States, United Kingdom, and France, different factions within the Soviet Union championed either crushing all German industry, on the one hand, or rebuilding a German economy that could be useful for exploitation, on the other. In the view of Sir Alexander "Alec" Cairncross, a British economist involved in quadripartite diplomacy during this period, occupation policy in particular became a feud between two factions: one led by Andrei Zhdanov and Nikolai Voznesensky (high-level officials in the Communist Party of the Soviet Union) and another informal coalition of Georgy Malenkov and Lavrentiy Beria (the chairman of the Council of Ministers and head of the NKVD, the predecessor to the KGB, respectively).[7] In 1943, Malenkov and Beria were appointed to the Committee for the Rehabilitation of the Economy of Liberated Areas, and in 1944 Malenkov became chairman of a special committee responsible for dismantling German industry.

Malenkov, in particular, was extremely influential in driving a policy of crippling German recovery through dismantling plants and, when possible, shipping them to the Soviet Union.[8] In this way, Malenkov's goal was much like that of US Secretary of the Treasury Henry Morgenthau. Malenkov's plans to cripple German recovery (like Morgenthau's) soon faced harsh criticism from those who saw such dismantling as costly and inefficient. The Zhdanov faction leapt on such stories to push for an end to dismantling. Continual reorganization of the occupation's bureaucracy followed, dictated at least as much by politics within Moscow as by anything directly to do with the Soviet zone of Germany. Throughout this in-fighting, much of the day-to-day operation of the Soviet Zone was left to the Soviet Military Government's (SVAG) propaganda department, led by Colonel Sergei Ivanovich Tiulpanov.

It is not my intent to provide a thorough breakdown of these internal Soviet policies in this book. In the context of this book, however, the lack of a clear, central authority had major implications for taking German military and industrial science and technology. American and British military governments certainly struggled to contain FIAT and related programs, even when they saw them as damaging the zonal economy, but they had more influence than their Soviet counterparts.[9] The looting of German industry that followed was not only far more ambitious in the East than in the West, but it was also far less organized. The resulting dismantling devastated the East German economy in ways still visible today.

Looting Germany: From War Booty to Industrial Dismantling

Especially in the early months, Soviet forces took enormous amounts of reparations from Germany. At the level of individual soldiers, looting "war booty" was common across each of the armies. It was especially pronounced among Soviet soldiers, many of whom were shocked by the decadence of German society relative to the material poverty of rural Soviet territories.[10] More formally, privileged reparations gangs, or "trophy teams," moved right behind the front lines. These gangs had officers' ranks, meaning local commanders had little ability to dispute their orders. This left these commanders in a difficult position when faced with reparations gangs representing different departments demanding the same materials.[11]

These teams were especially active through August 1945, when the free-for-all of the early seizures nominally gave way to a formalized reparations infrastructure, though industrial dismantling and removals certainly continued well into the 1950s. These trophy units were assigned large numbers of a wide variety of goods: four thousand trucks, thirty thousand cattle, five thousand pianos, fifteen hundred accordions, and so forth.[12] The thought that went into these wish lists is unclear but seems to have little to do with what was necessarily available in Germany. Meanwhile, both Soviet soldiers and reparations teams faced competition from another large group: those who had been enslaved by the Nazis and now wanted to go home, often to Poland, and grabbed what they could from (and often committing violence against) their former tormenters along the way.[13]

Beyond finished goods, Soviet reparations teams focused heavily on seizing German industrial technology. This was not a case of "intellectual reparations" in the sense that I emphasize in the rest of the book. Still, the Soviet Union undertook these large-scale industrial seizures far more than the other Allied powers, and ultimately they constituted another method of taking German technology back to the Soviet Union. Reparations teams would survey the zone, marking factories needed in the Soviet Union. They (or the occupation government) would dismantle the factories and pack them up, and the factories would travel by rail into the Soviet Union. There, the industrial department in need would receive, rebuild, and utilize the equipment. Out of about 17,000 "industrial objects" (most of which were large or midsize factories) that Soviet reparations teams surveyed in their initial trips through Germany in mid-1945, they marked about 4,300 (25 percent) for removal into the USSR.[14]

Despite Allied bombing and looting, there was still much to gain from both intellectual and physical reparations in what became East Germany. As Raymond Stokes argues: "In terms of physical plant, the Soviet zone was relatively no worse off, and probably far better off, than the western zones as the war came to an end," and even the zero-sum game of reparations seizures was "not necessarily doing irreparable harm to the East Germans by removing irrelevant, worn-out, or obsolescent technologies."[15] There were many important "high-tech" industries in this area, such as the famed precision optics enterprise Carl Zeiss in Jena. American and British teams took documents and workers from many of these plants with them when they handed over this territory to Soviet control, but much of what mattered—including key personnel—remained in the area. In the case of employees, offers of employment and stable rations were enough to entice some to return during the immediate postwar years of austerity and near starvation.

The Potsdam Agreement gave the Soviet Union claim to 25 percent of any factories, industrial equipment, or other "productive assets" taken as reparations in the Western zones, and initial planning included the United States, the United Kingdom, and France sending finished goods from German factories to the Soviet Union. As diplomatic relations soured in the late 1940s, it became clear that the Western Allies were unlikely to follow through on early general agreements to dismantle factories in large numbers and ship them east, so Soviet planners turned to extracting everything possible from their own zone. This turned out to be well over one-third of the entire productive capacity of the Soviet occupation zone in Germany.[16]

This dismantling took place in waves, with a major one occurring in the months leading up to the August 1945 Potsdam conference. Looking back, one of the British negotiators on reparations, Alec Cairncross, saw this as no coincidence: "In the months before the Potsdam Agreement this went on at a breakneck speed, apparently from fear that the Agreement might set limits to what could be taken in reparations. The result was a great deal of destruction, with relatively little economic benefit to the USSR."[17] In May and June 1945, about 460 factories in the Berlin area alone were taken apart, ranging from coal mines to locomotive building plants, electrical works to railway repair shops.[18] Moving toward autumn, this expanded to include removal of railway tracks themselves and hundreds more factories, namely, those in papermaking, sugar-refining, brick-making, textiles, and others outside of heavy industry.

The haste, disorganization, and enormous scale of these industrial reparations led to incredible waste—at least in most accounts. In the words of one Soviet administrator there at the time, "the dismantling of German industry . . . was characterized mainly by the almost complete absence of overall direction, particularly with regard to the technical questions involved in dismantling complicated industrial equipment."[19] Reports from the period regularly note equipment broken during dismantling or travel and trainloads of equipment rusting in depots in Germany or the Soviet Union.[20] What did arrive in decent condition suffered from another problem: a lack of both written information and know-how needed to reassemble and use the equipment. As one participant reported: "When the equipment came to be assembled, blueprints and layouts were often missing because the dismantling crews made a bonfire of all the paper in factory offices."[21]

The dismantling devastated the morale of German workers, including those otherwise ideologically friendly to the Soviets. This was partly because Germans were forced to participate in dismantling their own factories, in many cases working twelve to fifteen hours per day, seven days per week, with no pay.[22] According to Vladimir Rudolph, "there was also the attitude, 'If we can't ship it out, it's better to destroy it, so that the Germans won't have it,' as Special Committee Representative Saburov put it at a meeting of the ministry representatives in Neuenhagen on July 2, 1945."[23] The opinions of these German civilians were hardly a high priority for Soviet authorities, particularly in the early months of the occupation. In the context of the exploitation of

German science and technology, however, this was another factor that undermined scientists' willingness to collaborate with the Soviet regime (in cases where they had any choice).

It is worth noting that not all historians agree with this assessment. In his 1973 study *Western Technology and Soviet Economic Development*, Antony Sutton argues that Soviet technological reparations were actually much more efficient than what the United States, the United Kingdom, or France attempted. Soviet policy, he suggests, focused on easily transportable and broadly usable items such as machine tools and equipment, which they fit relatively efficiently into existing Soviet production systems. German equipment remained in use in Soviet industry for decades, and the "backwardness in control instrumentation and computers [was due in part to] the technical nature of the transfers from the German electrical industry at the end of World War II."[24] Matthias Uhl similarly portrays a much less chaotic dismantling process in his study of the Soviet missile program, though he emphasizes—like many other historians—that these German contributions are a contributing side story to Soviet developments rather than a driving force.[25] We should take such stories in context, then. Some amount of this talk of Soviet waste might well be the result of early Cold War propaganda.

Military Technology—Rockets, Nuclear Weapons, and Other Priorities

As eager as Soviet reparations teams were to take German industrial technology, they were even more focused on a number of specific military technologies. As was the case for the Western Allies, the most important targets in Soviet eyes were rocket technology (specifically, the V-2 series) and any insights into the construction of atomic weapons. Additionally, like the other Allies, these targets accounted for a large percentage of the German scientists and technicians brought back to the home countries. In some cases, this was voluntary, or at least semivoluntary. An example is Manfred von Ardenne, a physicist who agreed to take a two-week trip to the Soviet Union to discuss atomic research—a trip that turned into ten years before he was allowed to return.[26] Others were given no such "choice."

The participants in the Soviet quest to master the V-2 ran into an early problem: American sabotage. American forces were the first to reach some of the most important V-2 production facilities in northeastern Germany, such as the main design/testing facility at Peenemünde (where Wernher von Braun's team worked). When it became clear that quadripartite agreements would assign these lands to the Red Army, US troops seized and destroyed what they could to deny the technology to the Soviets.[27] These seizures helped American production and somewhat hindered Soviet efforts to learn rocketry. Ultimately, however, there was just too much of value for the Americans to take or destroy.

Soviet managers and experts quickly moved into these former Nazi rocketry and aviation facilities, including the BMW jet engine factories and the famous Mittelwerk V-2 production facilities in Nordhausen. Once there, they gathered the items that remained as well as any of the local Germans who had been part of the design or production process.[28] More than simply restarting these facilities, the Soviets expanded them. Leading Soviet scientists moved in to establish entire research facilities (e.g., the Institute Rabe and later Institute Nordhausen), aiming to master and then surpass Nazi rocketry and ballistics research. Such facilities generally had Soviet military managers, but the workforce included thousands of Germans, from former low-level technicians to top scientists. Helmut Gröttrup, formerly a leading scientist at the Peenemünde, is a key example.[29]

There is little question that artifacts and raw materials taken from Germany formed the basis for the Soviet rocket program.[30] Key Soviet specialists, including Sergei Pavlovich Korolev, Valentin Petrovich Glushko, Boris Evseyevich Chertok, and Alexei Mikhailovich Isayev, spent months in these German facilities. There, they pursued German know-how as well as designs. In the words of historian Norman Naimark: "The Soviets' system for garnering German rocket technology was very different from the Americans'. Soviet specialists . . . immersed themselves not just in German technological innovations, but also in the German methods and organization of rocket production."[31]

While the German army had developed a significant production line for V-2 missiles, eventually launching more than three thousand missiles at Allied targets, there was never a significant Nazi effort to build an atomic bomb. Certainly, German scientists made important theoretical contributions before the war, but a combination of the few resources being available during the war and incorrect estimates for how difficult a bomb project would be undermined any serious investment.[32] This meant that Soviet teams sent by the Soviet atomic project (managed by the NKVD under Lavrentiy Beria) had no established facilities like Nordhausen to occupy and instead planned from the start to use German resources to build up labs in Soviet territory.

The American Alsos Mission was just one of the American initiatives that seized as many atomic scientists and related facilities as possible, in large part to deny them to the Soviets. This meant, for example, that the Kaiser Wilhelm Institute for Physics—and its director, Werner Heisenberg—was already in Western hands when the Soviets began hunting for expertise. However, the low-temperature physics laboratory led by Ludwig Bewilogua remained in Soviet territory, and Bewilogua had worked on an experimental uranium pile. This lab was broken down and packed up—including the cabinets, door-knobs, and washbowls—and shipped east.[33] The same was true for remaining parts of the Kaiser Wilhelm Institutes for Biology, Biochemistry, Chemistry, Anthropology, and Silicate Research. In these scientific institutes, as in the industrial laboratories, much of this equipment ended up broken, lost, or split up after being requisitioned by different labs.

Work in the Soviet Union was an attractive option for many scientists, especially in the early days when conditions were at their worst in Germany and information about conditions in the Soviet Union were most scarce. Some were drawn in by the Soviets' willingness to accept even the most ardent Nazis if they had scientific value (the United States did import scientists who had been active Nazi party members in some cases, but officially refused to do so, and did indeed turn away many German scientists for this reason).[34] A particularly clear example is Peter Adolf Thiessen, a German physical chem-ist and head of the Kaiser Wilhelm Institute for Physical Chemistry and Elec-trochemistry. Thiessen joined the Nazi party in 1925, before Hitler's rise to power, and received several awards for service to the party. Despite that, So-viet authorities were happy to classify him as having a clean past and to move him, his lab, and the labs of several other prominent scientists with whom he was in contact. Among them was head of Siemens Laboratory and Nobel laureate Gustav Hertz. Thiessen, Hertz, and other German scientists helped consult on uranium enrichment in the Soviet atomic program.

Tallying the German researchers' contribution to the Soviet atomic bomb project is difficult. In large part, that is because so many of the participants had reason to make their contribution seem either enormously important or completely negligible. The researchers themselves eventually returned to East Germany—as is discussed shortly—and once there, many wrote memoirs that emphasized their own importance.[35] Soviet administrators managing these scientists at the time, meanwhile, were operating during the height of Stalin-ist paranoia, and reliance on researchers who had worked for the Nazis was a political liability. They, and Soviet scientists cooperating and competing with them, had incentive to downplay the Germans' contributions. In 1951, Soviet Minister of Armaments Dmitrii Ustinov admitted that "the Germans working in the area of reactive technology rendered significant help to re-create and reconstruct German designs, especially in the first period of time." However, "owing to the long isolation from modern science and technology, the work of the German specialists has become less effective, and at the present time when principally new and more modern models of [rockets] are being created, they would not be able to provide significant help."[36] As historian Asif Siddiqi has shown, this sort of claim was part of a broader effort to minimize these German scientists' contributions.[37]

Ultimately, German scientists seem to have contributed in relatively minor but meaningfully ways to the Soviet atomic bomb project. Within Soviet territory, several of the German scientists had their own laboratories, including Hertz and Manfred von Ardenne in Sinop, and they worked on the problem of uranium separation from a number of angles. For the most part, they worked independently within these labs, reporting to the Scientific-Technical Council of the First Chief Directorate (Pervoje Glavnoje Upravlenije) in Moscow. Soviet teams collaborated with these labs but also developed their own processes in parallel, and Soviet scientists in general were already quite capable. David Holloway, in his history of the Soviet atomic bomb project, concluded that "the German contribution to the atomic bomb was small and limited," perhaps on the order of months saved.[38]

Operation Osoaviakhim and the Mass Seizure of Scientists

By late 1946, the first major wave of dismantling and seizure of industrial and academic research facilities was dying down. By this point, diplomatic relations between the Soviet Union and the Western Allies had broken down. General Lucius Clay, the commander of the American zone, stopped transferring supplies and reparations to the Soviet zone in May 1946 in response to the Soviets not sending food from the more agricultural east, and in turn the Soviets began a large-scale propaganda campaign against all US efforts. American forces stopped all dismantling of plants in July until a new quadripartite "Level of Industry" agreement could be reached, which was now very unlikely. In July, US Secretary of State James F. Byrnes attended the Council of Foreign Ministers meeting in Paris, where he proposed merging the American and British occupation zones. In October, the military governments of these American and British zones made another major push to end FIAT and BIOS, which had been in operation for well over a year (and CIOS still longer before them).

Historian Filip Slaveski sees this period (late 1946–1947) as one in which overall Soviet policy pivoted from looting Germany under the assumption that it would be left to its own devices afterward, toward controlling and re-building Eastern Germany under the assumption that the occupation would continue long term.[39] Thus, the original plan of turning the Soviet zone into an industrial wasteland—it had seemed like an assurance that a new German state could not become a military threat—now seemed like a liability. Soviet reparations teams removed about 210 additional industrial facilities from Germany in mid-1946, but this was far lower than in the earlier period.[40] As it turns out, this was the calm before a new storm of seizures, this time aimed directly at stocking Soviet labs with German expertise and equipment on a scale beyond anything attempted previously.

Probably the most dramatic event in the exploitation of German science and technology took place in late October 1946, when Soviet authorities en-acted Operation Osoaviakhim. On the evening of October 22, after weeks of quiet preparation, Soviet troops and NKVD teams rounded up about three thousand German scientists, engineers, craftsmen, and other technical spe-cialists, along with their families and possessions, and placed them on trains heading east.[41] No explanations were given, nor were excuses or objections allowed. Some were asked to sign contracts ahead of time, but most learned of their fate by way of a late-night knock on their doors. The scale of Osoa-viakhim is stunning. Nearly every major firm in war-related industries were impacted, including Carl Zeiss in Jena, BMW Stassfurt, Leuna, Siebel Works, Junkers, and Schott.[42]

Among the individuals taken were many of those mentioned in V-2 and atomic research, including Helmut Gröttrup. He and two hundred of his col-leagues working at the Mittelwerk rocket research facility were invited to a party with a Soviet general, after which they were informed that they and their families would be on trains to facilities in the Soviet Union the following morning.[43] After these rocket research facilities in Germany were emptied of contents and staff, Soviet teams destroyed them.

Even facilities already dismantled or looted faced seizure, and their staff faced involuntary removal. The firm Carl Zeiss had a long-standing reputa-tion for making precision optics, which were useful in scientific research as well as military equipment, long before the Second World War. The Zeiss plant in Jena was looted by American and British T-Force troops in the initial invasion of Germany, then by dismantling teams in the early years of Soviet occupation. In Osoaviakhim, 270 of the remaining technical staff found them-selves rounded up and forced to head east. The contracts these scientists and technicians were forced to sign stipulated three-to-five-year terms, though those working on topics of national security (e.g., rocketry and especially nuclear weapons) would have no opportunity to leave for at least a decade.

Why Soviet leaders in Moscow saw such a sudden, extreme move as necessary this (relatively) late in the occupation is difficult to answer definitively. It is certainly true that American and British intelligence agencies had been recruiting specialists within the Soviet occupation zone for years. American intelligence officers at the time worried that the event was a prelude to war. After a few weeks of things settling, they revised this assessment, concluding that the Soviets had been concerned with accusations that they had not held up their end of the task of deindustrializing Germany. Their use of German research facilities, now stocked with Soviet scientists in leadership roles, was technically a violation of quadripartite agreements. British planners had indeed (hypocritically) been planning a propaganda campaign against Soviet use of German war-related research, so it is possible that Soviet intelligence had caught wind of this campaign and hoped to act first.[44] The Soviet Union also had dire need of skilled industrial workers, scientists, and engineers—not least in order to install the plants that had been dismantled in East Germany.

Osoaviakhim was a propaganda disaster for the Soviet Union, especially as many technicians had just begun to hope that their livelihoods were in less danger of being dismantled through reparations seizures. Newspapers around the world reported on the event, often exaggerating its scale even beyond its astounding reality and writing in violence that was more often threatened than employed. The *Chicago Daily Tribune*'s banner headline article reported that 150,000 "German slaves" had been seized via these "Red Kidnapping Raids."[45]

The head of the Soviet occupation government, Vasily Sokolovsky, is said to have reprimanded the American military government for criticizing a behavior they had taken part in themselves, arguing: "I am not asking the Americans and British at what hour of the day or night they took their technicians. Why are you so concerned about the hour at which I took mine?"[46] Whatever right the American government had to criticize such seizures, the press coverage and relatively late date for such actions made an impression on scientists. American, British, French, and Soviet intelligence agencies continued a game of attempting to recruit scientists from each other's zones well after Osoaviakhim, and this was now a much harder sell. When asked why few Western scientists would agree to take faculty jobs in East Germany, Werner Heisenberg explained through parable: "the fox sees many trails leading into the bear's cave, and none coming out."[47]

Long-Term Mechanisms for Exploitation in Place

Soviet policy toward controlling science and technology in its occupation zone was interwoven with other, overriding policy goals: reparations, negotiations with the Western powers over a potential German successor state, industrial policy, and establishing a socialist society in the Soviet zone, among others.

Soviet dismantling and seizure of German factories and equipment came in waves, but dismantling generally gave way to establishing infrastructure for longer-term control over German industry and science after Osoaviakhim. There were exceptions—in the winter of 1946–1947, another wave of seizures focused on power stations, printing works, and weapons factories (which theoretically should have been shuttered long ago but had been kept in production to meet Soviet demands).[48] By this point, such reparations could be compared to getting blood from a stone. Over 80 percent of the machine tool capacity and 60 percent of the light and specialized production tools were already gone. German workers' morale was also devastatingly low, which was another reason productivity had dropped to 50–75 percent of prewar levels.[49] With the possibility of a long-term East German state becoming more realistic, Soviet authorities saw more value in sustaining industry there.

As a middle ground, the Soviet occupation authority took ownership of several of the biggest industrial facilities, which became known as Sowjetische Aktiengesellschaften (SAGs).[50] These SAGs retained much of their original skilled workforce, sustaining some of Germany's technological traditions and providing employment for workers who might otherwise have fled to the Western occupation zones (or abroad). In most cases, the Ministry of Foreign Trade owned 40 percent of each SAG, with the remaining 60 percent owned by a relevant industrial ministry or other enterprise.

This move toward establishing SAGs started shortly before Osoaviakhim and suffered from both physical and intellectual reparations programs. The first SAG, established in January 1946 to assure the delivery of scarce concrete to the Soviet Union, suffered from the inability of a dismantled electrical parts industry to supply needed repair and upgrade parts. By April 1946, about fifty to sixty of the biggest industrial plants marked for reparations were instead converted into SAGs. By December 1946 (two months after Osoaviakhim), 30 percent of all production in East Germany had been converted to SAGs. The SAGs were particularly prominent in security-related industries, including the large uranium mines on the border of the Czech Republic that became Wismut SAG in 1947. Despite any security concerns, the Soviet administrators were not generally bothered by rehiring former Nazi party members, providing further continuity with prewar German industrial traditions.[51]

Like the overall reparations programs, decisions about selecting, expanding, and eventually dissolving SAGs depended in large part on the internal politics of Soviet industrial ministries. Some were returned to German control even in 1947. By the end of that year, SAGs controlled only 25 percent of production in the Soviet zone, down from 30 percent the previous year.[52] However, most remained under Soviet control through the establishment of a relatively autonomous East German state in the 1950s. Wismut SAG remained under partial Soviet control much longer.

Though SAGs were not implemented with technology transfer as a top priority, they served as stations where Soviet engineers and technicians could collaborate closely with skilled German workers for long periods. Conversely, as Joachim Radkau argues, SAGs "anchored . . . structures of Russian applications of technology in the production apparatus of the GDR"—meaning, among other things, that they could impose on East Germany a Soviet engineering tendency to design grand-scale technologies without much regard for efficiency.[53] Since the Soviet Union preferred to work within the Soviet bloc as much as possible (not least because of shortages of hard currency for trade with the West), East German enterprises eventually became comfortable producing goods for this semi-captive Soviet market, even when they were not competitive in international markets. Raymond Stokes argues that "the Sovietization of GDR technological culture was slight," at least through the 1950s, but the institution of these SAGs was a step toward this longer-term trend.[54] Eventually, though SAGs were in part designed to funnel German techniques to Soviet industry, they facilitated flows of technology in both directions.

While the SAGs forced long-term cooperation between German technical workers and Soviet counterparts and management, Science and Technology Offices (*Nauchno-teknicheskii otdel*, or NTOs) coordinated scientific research in the Soviet zone. Just a month before Osoaviakhim, the Soviet Council of Ministers approved the Administration for the Study of Science and Technology in Germany to be attached to the occupation government.[55] This group, in turn, supervised NTOs throughout the Soviet zone. By the end of 1948, thirty-six NTOs employed 611 Soviet specialists who oversaw six thousand German scientists and seven thousand German workers.[56] They worked on a mix of industrial and military research, sponsored by different Soviet departments. Examples include the contracts for research into aviation, polymers and plastics, textiles, and meat and dairy production.

Even more directly than SAGs, NTOs were a major venue for German industrial technology and know-how to pass into Soviet industry. These NTOs, as well as their teams of German specialists, were vital in the formation of a variety of Soviet industries, especially in chemicals. High-octane gas, carbon fuels, turbines for liquid fuels, nylon, coal briquettes, and synthetic rubber are important examples.[57] Further, Soviet ceramics, metal-finishing, film-developing, and metal-plating industries benefited majority from their use of NTO research contracts.[58] However, in some ways the NTOs' success undermined this technology transfer role. In January 1949, the Administration for the Study of Science and Technology in Germany conducted a study of problems facing the NTOs, and a primary finding was that too few Soviet specialists supervised too many Germans. This meant little time in sustained contact and transfer of valuable German know-how.[59]

Impact of Returning Technical Specialists in East Germany

In contrast to the German scientists brought to the United States who sought from the start to form a new, permanent community there, almost all of the German scientists taken to the Soviet Union returned to Germany within a decade of their removal. Some (estimated at around 25 percent)[60] quickly emigrated to West Germany or other Western nations, but the large majority remained in East Germany. There, given special status and guaranteed employment (and, for some, the chance to pursue a genuine ideological belief in building a socialist state), these returnees had a major effect on East German society.

The first set of these technical specialists to return arrived in September 1950, meaning those taken in Osoaviakhim spent at least half a decade under direct Soviet control. This first set of returnees arrived in several waves over the next few months, until, by the end of 1950, about 610 specialists (and 1,080 family members) had arrived in a resettlement camp at Wolfen, a town about twenty-five miles (forty kilometers) north of Leipzig.[61] Altogether, this represented about 25 percent of those taken to the USSR.

The remaining specialists returned in waves, dragging along deep into the 1950s. Another 10 percent returned in 1951, then another 31.2 percent in 1952.[62] By 1954, over 90 percent had returned, but the final 10 percent were in some ways the most high-value of all. German researchers who had been working on atomic weapons, rocketry, and in other key military industries were held longer in the Soviet Union as a kind of "cooling off" period, with the assumption that within a few years their knowledge would be too outdated to be especially damaging. In all, by the end of 1958, about 2,500 German scientists, engineers, skilled craftsmen, and other technical workers returned from the Soviet Union.

The facilities in the resettlement camp in East Germany were terrible, with inadequate sanitation and several families living in each room. Still, it was temporary: the East German government was keen to make good use of this skilled manpower and set out to find productive jobs for each returnee as quickly as possible. The Soviets had chosen these specialists for their scientific and technical prowess, and East Germany had ambitions of being a scientific powerhouse.[63] As historian Andre Steiner has pointed out, "success in getting these prominent scientists and engineers to commit themselves to the GDR provided the country with urgently needed prestige, and . . . with a certain degree of legitimation."[64]

In addition, much like the American government pursued policies aimed at denying scientific manpower to the Soviet Union, the Soviet Union was concerned about what information these returnees might provide to the West. This was especially true, of course, regarding atomic research. In March 1955, when the first group of these nuclear researchers was allowed to return, Soviet authorities provided suggestions on keeping them in the USSR. Manfred von Ardenne, they suggested, should be kept in East Germany "by paying attention to his avarice and his need for admiration."[65] Those who had freely admitted their intentions to head West found themselves held in the Soviet Union for some time longer.

The Soviets (and East Germans) were not paranoid to think the American and British intelligence agencies were actively seeking out these specialists. As Paul Maddrell details in *Spying on Science*, the Western powers surveyed all refugees and immigrants from the eastern bloc for those with experience in science and technology—even just working in a uranium mine or laboring in the construction of new plants.[66] Beyond this, they employed agents to induce defection among key scientists by offering high salaries and other perks in the West. The goals here, as with the wartime T-Force efforts to secure German scientists, were twofold: gain scientific manpower for the West and, even when such scientists were not needed there, slow scientific progress in the East by denying them this resource. In the context of the early Cold War period, governments increasingly treated scientific manpower like a commodity similar to uranium deposits: a target to be controlled, hoarded, and smuggled in a zero-sum game.[67]

For the East German economy, these returnees represented a skilled workforce far beyond academic scientific research. Only 8.6 percent had doctorates, and less than 1 percent became professors.[68] Indeed, 45 percent had no university training at all. Some of these individuals were likely swept up in the indiscriminate nature of Osoaviakhim and related forceful Soviet tactics, but most were specialized craftsmen, such as skilled welders or machine tools specialists.

Still, placing these workers was a daunting task. As the minister of labor and health services for the Saxony-Anhalt region commented: "It is very difficult to place some of these individuals, since they are very specialized."[69] Worse, many were taken to the Soviet Union in the first place because their specialty lay in industries now forbidden in occupied Germany, such as aircraft production and military research. Some were placed in positions in related industries, such as shipbuilding, but much of their expertise was lost in moving countries, production styles, and now industries. The Soviet decision in 1954 to allow a revival of the East German aircraft construction industry alleviated this problem, and returnees were central to that industry in the years to come.[70]

Those who found jobs were greeted as "nobility without titles," both as a continuation of proud industrial tradition and as workers who had sacrificed to rebuild the Soviet Union.[71] None of the occupying powers ultimately followed through completely on denazification of industry or academia (and would have had a hard time doing so without leaving a large portion of the population permanently unemployed and many important positions unfilled), so these returnees' earlier affiliations were not much of an obstacle. As historian Dolores Augustine has chronicled: "Nazi-era scientists had brilliant careers in East German industry, were accorded many privileges, and enjoyed the confidence of the Communist leadership to a surprising extent."[72]

Conclusion

Soviet reparations policy—including that which focused on science and technology—was ruthless, and its tremendous scale often made it ineffective. Still, for some Western observers, this seemed like a pragmatic and efficient Soviet government seizing an opportunity missed by the other Allies. As Peter Nettl wrote in *Foreign Affairs* in 1951: "Only the Germans and the Russians have so far managed to absorb large-scale reparations successfully. . . . There is no magic which distinguishes Soviet success in extracting reparations from Germany from Western failure, unless there is magic in the commonplace that if you really want something very much, you put it ahead of other things. The Russians . . . have sacrificed Allied goodwill, they have sacrificed the German Communist Party, and they have alienated German public opinion."[73]

This perception that the Soviets had done a good job of getting value from reparations actually had a major influence on Cold War diplomacy. The heavy-handed Soviet seizure of German scientists played into Western prejudices about the Soviet Union, at least until the launch of Sputnik in 1957. In the words of General John B. Medaris, many US military and government officials considered the Russians a "retarded folk who depended mainly on a few captured German scientists for their achievements, if any. And since the cream of the German planners had surrendered to the Americans, so the argument ran, there was nothing to worry about."[74] This stark underestimation of Soviet scientific capabilities not only led to overconfidence in the American ability to sustain a monopoly on nuclear weapons but fed paranoia about spies smuggling scientific "secrets," since that seemed to be the way the Soviets could get ahead. The underestimation also led to a panicked overestimation after the shock of Sputnik, including the hysteria around a nonexistent "missile gap" and "bomber gap."

In terms of actual influence on Soviet society, there seem to have been real gains, but it is important to avoid the aforementioned idea: that Germans were primarily responsible for Soviet successes in science and technology. German scientists contributed to the Soviet atomic project, but Stalinist paranoia and the competence of Soviet science without them meant that they were kept at arm's length. A similar story seems to be true in rocketry/space research, and to a lesser extent in other industrial areas. With these caveats in mind, German scientists contributed in each of these areas, speeding up research and helping to avoid costly diversions down the wrong path—probably more so than in the other Allied nations, who started from a stronger industrial base.

Dollar-value accounting, of course, is even less realistic here than in the case of the Western market economies, due to the nature of the Soviet command economy and dense, opaque political structure. The USSR undoubtedly took a more aggressive stance toward reparations seizures of all kinds, including intellectual reparations, and the cost to buy those same materials elsewhere would have been high. What about science and technology in the broader sense, though? The USSR spent heavily, buying nearly every FIAT and

BIOS report, but the very structure of intellectual property law operated much differently in communist states than in Western ones. One NTO chief even complained that this was a reason for confusion about who owned the fruits of NTO research: "We need patent laws; we need patent offices."[75] The Soviet Union and its satellite states tended to offer one-time rewards for inventions, but inventors were required (or at least heavily encouraged to make little difference) to turn over their innovations to the state.[76]

Despite these fundamental differences, there were striking parallels between Soviet exploitation of German science and technology, and that in the other occupation zones. One major comparison is the basic chaos, duplication, and internal competition in the process. Like in the United States, the United Kingdom, and France, a variety of Soviet administrators sent investigative teams and fought over who benefited first. Though more pronounced than in other zones, SVAG, like its American and British counterparts, resisted these intellectual reparations programs as undermining the zonal economy, thereby making their job more difficult. Like these other zones, SVAG's protests were mostly ignored, especially before late 1946.

Like the French, Soviet planners emphasized capturing entire teams of scientists and technicians, rather than hiring individuals piecemeal. They were more willing and able than the French to transplant entire factories and research institutions, but they certainly did not quibble about the importance of grabbing even the lower-ranked factory workers and laboratory staff. Though they did not discuss it in terms of know-how, Soviet control of German research facilities similarly focused on sustained, long-term exposure to processes, from start to finish.

The highest priority for Soviet investigators was the same as for the other Allies: learning how to master and develop beyond V-2 rockets, and to discover as much as possible about German atomic research. Here, the enormous resources that the Soviets threw at intellectual reparations paid off. Though atomic spies such as Klaus Fuchs likely contributed more to the Soviet atomic project than the teams of German researchers, they still offered valuable innovations and scientific manpower. In rocketry, reactivation (and later seizure) of V-2 production and design facilities provided valuable insights that Soviet scientists were ready to build upon.

In a sense, the longer-term results for East Germany were a kind of twisted version of what some historians have argued for West Germany: the intellectual reparations programs built familiarity and served as a kind of conveyor belt of business relations for the postwar decades. Again, though, this was more literal and forceful in the Soviet case, as SAGs, NTOs, and the general structure of Soviet bloc economics ensured sustained technology transfer and economic trade. East German technical expertise and know-how remained the envy of Eastern Europe for years, even if its innovation slowed relative to the West. Once again the reparations programs did not, in effect, "steal" technology in the sense of depriving its owner of its full use. As Raymond Stokes put it, the "GDR at its official founding was not all that far behind the Federal Republic in technological terms, was still competitive in key industries, and was still very good in science and engineering education and practice."[77]

In the 1950s, a number of factors began to eat away at the East German technological system and thus at its ability to deliver cutting-edge technology to the Soviet Union. Some factors were specific to the intellectual reparations program—specifically, one of the long-term consequences of Osoaviakhim was that East German scientists never felt fully secure about their position in society. Other causes were much broader: the Stalinist social system, the end of the illusion of democracy, and the general brain drain of skilled workers fleeing to West Germany before the construction of the Berlin Wall.[78] In order to get a better sense of what was at stake, we turn to one area where Germany excelled both before and after the war, and that sometimes even spanned this growing chasm between East and West: the world of academic science.

5

Academic Science and the Reconstruction of Germany

The same agencies involved in taking Nazi technology (FIAT, BIOS, etc.) were also responsible for deciding the fate of German science and scientists. To modern eyes, that can seem a little strange. Learning about German wood pulp production techniques and how to mass-produce specific textile dyes is a very different enterprise than deciding whether theoretical physics professors in elite universities had been too eager to join the Nazi party. For those planning the occupation of Germany, though, science and technology seemed fundamentally linked. Throughout the Allied nations, agencies staffed by scientists made claims to controlling both.

In part, this is a result of scientists' dramatic increase in political influence due to the war. Both nuclear weapons and radar were convincing arguments to many politicians that basic science deserved funding not only for its own merits but because it was a necessary and reliable way to develop weapons and economically valuable products. If the First World War had been influenced by poison gases and machine guns, the Second World War had been all the more decided by science-based technologies. The next major war, it stood to reason, would rely even more on science. At least listening to scientists' ideas on how to pursue that science seemed prudent.

In this chapter, I step away from the purely applied world of industrial technology to look at how "basic" (as opposed to "applied") science influenced the reconstruction of Germany.[1] Within each Allied nation, debates raged about how best to sponsor, harness, and protect science at home, and for each, these debates spilled over into their thinking about rebuilding Germany. Should scientists be key advisors to presidents and prime ministers? Was science fundamentally apolitical, potentially corrupting, or did its much-touted openness and internationalism make it an actively positive influence on the German people? Should decision-making about science funding be left to scientists, given over to military planners, or even opened up to democratic decision-making? How could science best be mobilized for economic growth? By simply dumping money into basic research or through directed investment in particular technologies? This swirling, evolving set of questions occupied officials in Washington, London, Paris, Moscow, and the occupation zones of Germany alike.[2]

Science policy in occupied Germany became closely tied to science policy at home for each of the occupying nations. For both those who wanted to crush Germany and those who wanted to rebuild it into a peaceful member of the international community, science and scientists were invaluable tools, in ways that reflected policymakers' changing ideas of what science could do for the nation. What this meant in each nation varied, of course.

For the United Kingdom, German science was far less a credible threat than an opportunity, provided German scientists could be enlisted in British economic interests. For the United States, science represented a source of "soft power" diplomacy. The newly founded American intelligence community joined with nominally private philanthropic groups such as the Ford and Rockefeller Foundations to rebuild Western European science, partly in order to build goodwill toward America that would be useful in other areas. For France and eventually both East and West Germany, science was a source of cultural prestige and legitimacy, both desperately needed after years of occupation. German scientists and policymakers were hardly passive subjects waiting to be molded by the occupying powers, and they had their own policy goals in mind throughout these debates.

One of the major impediments to creating any effective policy regarding Germany was the structure of the Allied Control Authority, whose component Allied Control Council required unanimity in every decision. Science fell under its Economics Directorate, then the Committee on the Liquidation of War Potential in Germany formed to craft Allied science control policy. Allied Control Law No. 25, passed in April 1946, dealt with "the control of scientific research" in order "to prohibit scientific research and its practical applications for military purposes, to control them in other fields in which they may create a war potential, and to direct them along peaceful [economic] lines."[3] It distinguished between "fundamental" and "applied" research and development, placing heavier restrictions on the latter, and between research of a "military nature," "peace-time applications," neither, or both.[4] War-related research was banned, peaceable research was allowed, and those in between were decided on a case-by-case basis; however, the definitions of these terms were nebulous. Enforcement was left to the individual military governors. While there was some early agreement about the danger of science left untended, its role in the development of postwar Germany varied substantially across the zones.

Even a nation-by-nation approach imposes too much artificial order, though it makes for useful comparisons. Within each nation, factions debated the best policies, thinking changed over time, and the quick evolution of the early Cold War continually upped the stakes for all involved. All of these complex, interacting forces (and many more beyond) shaped efforts to control, exploit, and sometimes encourage German science and technology.

German Science from Weimar to Occupation

Through most of the nineteenth century, Germany had a worldwide reputation for science and technology. The research university—with professors who split duties between teaching and research—originated in Germany, before eventually spreading to other nations. Since universities were major cultural institutions funded by the state, the successive German governments of the nineteenth and twentieth centuries (from empire to Weimer Republic to Nazi to divided Germany) had a major investment in science. Universities were not the only places where science research took place, of course—German industrial concerns invested heavily in research and development, especially in chemicals and other "high-tech" fields of the day. These, too, received substantial support from the state.[5]

By the early twentieth century, however, German scientists were concerned about a perceived loss of ground on the world stage. The United States was a growing powerhouse, fueled by a relatively enormous economy. Among those most benefiting from this wealth were industrialists such as Andrew Carnegie and Nelson Rockefeller, both of whom founded charitable institutions upon their deaths. Since the American government invested relatively little in science before at least the First World War, contributions from philanthropic groups such as the Carnegie and Rockefeller Foundations were central to American science funding in the late nineteenth to mid-twentieth centuries. These private foundations (and other philanthropists) had the advantages of not needing to appease taxpayers or play politics in the same way that a federally funded agency might. They could also potentially fund research institutes where scientists focus exclusively on their research, without the teaching and administrative responsibilities that universities imposed.[6]

These anxieties about German science's place in the world led to the formation of the Kaiser Wilhelm Society (Kaiser Wilhelm Gesellschaft, or KWG) in 1911. Following what they saw as the American model, the KWG was comprised of many subject-driven institutes, such as the Kaiser Wilhelm Institute for Brain Research. These institutes were separate from universities and funded by support from industry, state funds, and donations from private philanthropists. The KWG was extremely successful on the whole, drawing in researchers from around the globe and generating work that would lead to twenty-one Nobel Prizes (three of which were earned during the Nazi Third Reich, though the government refused to allow the researchers to accept the prize in protest of a Jew having won the Peace Prize).[7] Meanwhile, the dire needs of German science following the First World War led Max Planck, Fritz Haber, and Ernst von Harnack in 1920 to found the Emergency Association of German Science (Notgemeinschaft der Deutschen Wissenschaft), an umbrella organization comprised of universities, the KWG, and scientific academies. In 1929, this group was renamed the Deutsche Forschungsgemeinschaft (German Research Foundation).

Following Hitler's rise to power in the 1930s, one of the first laws passed began the systematic expulsion of "non-Aryans" (mostly meaning Jews), communists, and others deemed unfit from civil service, including from universities. This began an exodus of Jewish scientists, especially starting in 1936, that seriously damaged German science. Albert Einstein, Max Born, Fritz Haber, and many other former or future Nobel Prize winners were among the hundreds of scientists who fled or were expelled. This mass exodus was a major boon to British and American science (and, to a lesser degree, to the Soviet Union, France, and many other nations), and correspondingly hurt German science in both reality and prestige.

German scientists' responses to Nazism, similar to the responses of professionals and intellectuals throughout German society, ran the gamut. The most prominent of the ardent supporters were Nobel Prize winners Johannes Stark and Philipp Lenard, who promoted an ideologically driven *Deutsche Physik* (German physics). Many other scientists were content to support Nazi ambitions if it meant funding and stability. Most sought to retain as much independence as possible, continuing their research while having no real allegiance to (and, rarely, even actively resisting) the Nazis. The KWG, led by Max Planck, resisted Nazi control as much as possible while also generally avoiding any active resistance against the regime.[8] The Deutsche Forschungs-gemeinschaft, however, fell under the control of Johannes Stark and the Nazi government's interior ministry (and, later, the education ministry).[9]

In the immediate aftermath of the war, questions arose as to how to "denazify" German society, and once again universities (and other educational institutions) were at the forefront. The process of denazification was messy, controversial, and uneven within any zone, and inconsistent across zones. The goal seemed clear enough: to punish those who had committed war crimes and remove dedicated members of the Nazi party from positions of authority, including in classrooms. The problem, of course, was in figuring out each person's level of involvement: A dedicated true believer? Someone who had disliked the Nazis but went along with things for fear of punishment? Someone who had subverted Nazi goals, even passively (e.g., by working less capably than possible on war-related production)? Those who joined the Nazis before they seized power were assumed to be the worst of the worst, and they were easy to identify from captured records (and because they had usually been given positions of power in the Third Reich). The rest—the vast majority—were a mess of conflicting testimonies (nearly anyone could find someone to testify that he had been against the Nazis from the start, whatever the truth) and intractable moral questions about whether following orders and "just going along" was enough of a crime to merit being banished from public life. The occupiers were well aware that any denazification efforts were going to be flawed at best, but the attempt mattered in and of itself.

Beyond denazification for its own sake, the question of how to handle science raised a unique set of issues, in part because of different understandings within and among the Allied nations as to their own goals. Each of the Allied nations had voices championing positions along a spectrum between two main choices:

1. Destroying Germany's ability to ever wage war again. This included dismantling/destroying/taking industry, reeducating Germans, and possibly indefinitely occupying/annexing German-speaking lands into other nations (or into international control).

2. Rebuilding their own occupation zone (or possibly combined zones) in their own image, integrated into their broader sphere of influence. This involved rebuilding industry, reeducating Germans, and ensuring that their portion of Germany's economy was interdependent with the rest of this bloc.

As domestic politics changed, so, too, did priorities in Germany. American voters and policymakers grew more fearful of Soviet power than German resurgence. British concerns about preventing economic and diplomatic decline pushed them closer to the United States and toward pragmatic stances on ruling Germany. French communists and Gaullists battled over the legacy of the resistance. The Stalinist period of Soviet politics was full of infighting, purges, and growing concern with what they saw as Western aggressive imperialism, which would require both a stronger military and a buffer zone in Eastern Europe.

No one was interested in repeating the settlement of the First World War, and one of the mistakes many perceived from that recent history was allowing a powerful combination between universities, independent research centers (like the KWG), large-scale industries, and the military—what we might today call a "military-industrial-academic complex." Dismantling science might make for a more peaceful Germany. Conversely, throughout the course of the twentieth century, scientists around the world had self-consciously built up influence and prestige by promoting science as apolitical, international, open/transparent, and even actively moral.[10] Science, then, might be useful in reeducating and "denazifying" Germans. On-the-ground developments in Germany, politics back home, and international diplomacy all came together in the occupation zones to decide science's fate in occupied Germany.

For the Germans themselves, meanwhile, science was seen as part of *culture*, similar in some ways to film or literature. Even the German word for science, *Wissenschaft*, is much broader than the English term, encompassing systematic study of any field: linguistics, literature, physics, and so forth. One of the internal battles of rebuilding science in West Germany centered on whether the education or culture ministries should be responsible for sponsoring science. To the extent that Germans had influence on occupation policies (which varied over time and across zones), this was in some ways a question of what German culture would become in the wake of two crushing wars and an uncertain future.

British Leadership through Goodwill among Scientists

There was no singular "British plan" for dealing with Germany, and ongoing debates among (and within) different departments led to inconsistent and changing policies. To generalize, the British bureaucracy broke down into two camps when it came to industrial dismantling and reparations in Germany. On one side, the Foreign Office and Treasury were in favor of rebuilding Germany. On the other, the Board of Trade, the Ministry of Supply, Admiralty, and other supply-oriented groups advocated for rebuilding German industry in ways that would allow it to be used to Britain's short- and long-term interests.[11] Within the British zone of occupation, the Control Commission for Germany, British Element (CCG/BE) listened to each of these departments while employing its own staff to develop goals and policies.

For the CCG/BE's influential Science and Technology Research Board, the scientific threat from Germany was something to consider in terms of decades and generations rather than in months and years. This threat (if one existed) was more of a politically useful specter than a pressing reality, and without a fear of German science, they instead saw opportunities for using science to promote British goals.[12] It had become apparent to them since at least 1946 "that the potential threat to peace from failure to control fundamental or reasonably small scale applied research is a slender one. . . . The things of which we are afraid are: atomic energy, bacteriological warfare, guided missiles, chemical warfare, and the at-present unknown scientific advance which is going to produce the war winning weapon of the next war."[13] The first four items would require substantial engineering and industry, which would be simple to detect. The last (the unknown, e.g., the next war's development akin to nuclear weapons) was the most important threat but also one that they could not hope to see coming.

This might have been at least partially a reflection of the predominance of scientific researchers among the Science and Technology Research Board's staff, who saw the autonomy and prestige of science as being particularly worthy goals. The quadripartite law limiting German Research, Law No. 25, was "largely designed to combat a danger which does not exist," but the records it created about the organization of German research, and what other powers were investigating, "gives us records of real use . . . for defence intelligence purposes."[14] This, in turn, made it all the more important (in their own eyes) that the Science and Technology Research Board retain authority for enforcing the law. "Otherwise [enforcing the law] might be regarded as a function of the Intelligence organisation. That would not be desirable, because Intelligence organisations are not in general staffed by men of a type who are able to maintain good relations with high-grade German scientists . . . [and] we regard the maintenance of such relations as a cardinal point of our policy."[15]

This goodwill of German scientists was a lynchpin of British policy and more consistently a priority for them than either the American or Soviet authorities. In part this was tied to ambitions (discussed in chapter 2) to reverse a perceived flow of British science leading to German industrial products. It was also closely tied to concern about British prestige and place in the world, now that many colonial assets had been sold off and others (e.g., India) seemed likely to demand independence in payment for their war service. "It is general experience," one British occupation authority policy document from June 1948 reports, "that, of all the different communities within the nations, it is the scientists who tend to be most international in outlook and most able to co-operate closely with one another." If they could ensure the goodwill of these "high-grade German scientists . . . (provided they do not belong to a politically highly undesirable type)," then, for example, "we should further establish friendly relations with the French Control Commission."[16]

One clear example of the British occupying authorities actively pursuing the goodwill of German scientists was the reformation of the KWG as the Max Planck Gesellschaft (MPG) in 1946.[17] Though the KWG had retained substantial independence from Nazi control, some of its institutes had conducted war-related research. That was enough to cause the Americans to label the society as a tool of fascism and call for its dissolution, an argument that found sympathetic ears in France. The Soviet Union saw value in a centralized German research institute but preferred the German Academy of Sciences as a tool for molding the "German bourgeois intelligentsia" into good socialists.[18] The British resisted these efforts, though initially unsuccessfully. The inter-allied Allied Control Council passed a law that would dissolve the KWG on July 11, 1946, on justifications that it had performed war work and represented a kind of "research trust."[19]

The now-former KWG still had support from prominent German scientists, however, including recently elected society president Otto Hahn. Their lobbying convinced the head of the research branch of the British occupation authority, Colonel Bertie Blount, to continue working toward a reconstitution of the society. As Blount soon learned through advice from Sir Henry Dale, a scientist with many political connections, it was the name (Kaiser Wilhelm) and fact that it showed continuity from the Nazi era that upset the other Allies the most.[20] The refounded MPG, located in the British zone of occupation as of September 11, 1946, served as a potent draw for scientists from other zones.

From this position, the MPG was able to expand into the other Western occupation zones, though not without some resistance. As discussions commenced about how to integrate the MPG into a unified "Bizone" of the British and American zones, the Americans—by 1947 far more willing to rebuild

German science—sought to export their own model of science research by attaching former KWG labs to universities, à la the Princeton Center for Advanced Study or the Stanford Research Institute. The French minister of national education approved a similar plan in June 1946.[21] The Science and Technology Research Board of the British CCG/BE thought little of this idea, however, which they "would deplore as much as we should handing them over to industry."[22] This was because under such a plan, the KWG/MPG "would . . . be dependent upon the Education Controllers in the various Länder (states), whereas it was generally agreed that from the scientific point of view it would be better for them to be completely independent."[23] Science—and the good-will of the scientists, who wished to retain autonomy—would be the British policy. It was not without some active effort and cost, then—by German scientists foremost but backed by the British occupation authorities—that the MPG was refounded in the American and French zones on February 26, 1948.

It was not just currently trained, established scientists who drew British attention but also the next generations now training in schools and universities. Large numbers of students enrolled in the Technische Hochschule (universities focused on engineering) in the immediate postwar years: 6,383 in 1946–1947, up from 5,695 in 1934–1935. Overall enrollment in science and technology programs grew from 15,860 to 29,400 during the same period, an almost 90 percent increase. This created "a matter of serious consideration" about whether all of these students could be absorbed into the German economy—if not, they might be a source of scientific manpower open to international (including British) bidding.[24] While American plans often involved denying scientific manpower to the Soviets, for the British these students represented opportunities, if they could be convinced that the United Kingdom was their natural ally.

Planning turned to how to pursue most effectively "the desirability, which has been growing in importance, of ensuring that as many high-grade scientists as possible should come to the western zones, particularly the British Zone, and remain there." The answers were ones that would have been familiar to the American planners advising the Rockefeller Foundation to sponsor European science during this same era: providing facilities, copies of journals and other publications, travel funds for scientific exchanges and visits, the reestablishment of scientific organizations, and aid finding "proper" employment, either within Germany or abroad (within the Western countries, naturally).[25]

The British had a more old-fashioned empire to consider, and planning for dealing with German science took into account how best it could be used to benefit this empire.[26] The British CCG/BE was proud of its accomplishments in rebuilding German science, most notably through the formation of the MPG, but worried about who would benefit in the end from these efforts in the event of a Soviet invasion of the West. The ambition, unsurprisingly, was that the scientists of Britain would benefit from the "close association" forged with German scientific elites, but the Science and Technology Research Board worried that if the Russians overran West Germany too quickly, the Soviets would be the main beneficiaries. In response, the board suggested that over the next three to ten years, the British government create throughout the British Empire "a shadow organization of research institutes which could, if the situation deteriorates, be rapidly expanded at the expense of science in Germany."[27] The extent to which this was put into effect is uncertain, but the ambition of benefiting colonial science as well as industry through exploiting and supporting German science was clear.

American Conflicts of Science and Governance

The role of science and scientists in American democracy was an unsettled, hotly debated issue in the early aftermath of the Second World War.[28] Scientist-leaders such as Vannevar Bush gained prominence as head of the Office of Scientific Research and Development (OSRD) and pushed for permanent science advisors to presidents and other government officials. Groups of Manhattan Project veterans organized to lobby for international control of nuclear weapons. The creation of the National Science Foundation led to debates about whether scientists should choose what science to fund, or whether more democratic means should govern.[29]

In Germany, American policy swung perhaps the most drastically of any of the powers, from suppressing to supporting German economic and military power. This, in turn, meant changes in the prospects for scientists. During the initial invasion and occupation period, there was near unanimity that Germany should be prevented from ever becoming a threat again. This led to JCS 1067, a directive to General Dwight D. Eisenhower, the head of Supreme Headquarters, Allied Expeditionary Force. US Secretary of the Treasury Henry Morgenthau's "program to prevent Germany from starting World War III," outlined in JCS 1067, proposed to reduce Germany to an agricultural economy.[30] This meant stripping away Germany's industrial base and permanently suppressing any military output. Insofar as science was seen as generating the military potential, it, too, must be suppressed.

This shifted toward favoring rebuilding the (West) German economy—

and eventually even the German military—as part of an American-led coalition against perceived Soviet expansionism and aggression. The details of this policy shift (which were never complete or clear-cut) are mostly outside the scope of this discussion, and the subject of many excellent book-length studies.[31] Both viewpoints saw science as a vital tool for achieving their goals, however, and both left deep impressions on the shape of science in the American zone of occupation.

A key figure in the transformation of science policy in the American zone—and a representative of the influence of scientist-advisors—was Roger Adams, an organic chemist and head of the Department of Chemistry at the University of Illinois at Urbana-Champaign. In late 1944, President Franklin D. Roosevelt turned to Vannevar Bush, his science advisor and head of the OSRD, for suggestions on how best to handle science policy in Germany. Economists and prominent scientists had been using the election season to put pressure on FDR to soften American treatment of the German economy (and science), and FDR delegated this task to Bush. Bush turned to Frank Jewett, president of the National Academy of Sciences, who put forward Adams as an ideal candidate.

Adams took up the task of putting together a new strategy for German science, aided by a committee that included Isidor Isaac Rabi and other luminaries of industrial and academic science.[32] The committee's recommendations made little immediate impact in Washington, but the OSRD, War Department, and State Department jointly appointed Adams as "expert consultant" to the occupation government in Germany.[33] Adams received little welcome from General Lucius Clay, head of the Office of Military Government, United States (OMGUS), the military government in the US zone, who was not sure what to do with a science advisor. Adams set to work on his own, however, surveying and critiquing OMGUS's science policy. Nearly every agency within OMGUS seemed to be meddling in science policy, he argued, leading to conflicting regulations duplication and unclear overall goals. Only FIAT was well organized, and its mission was more concerned with American well-being than German.

Though newly arrived in Germany, Adams was appointed head of the quadripartite Committee for the Liquidation of Military Potential in November 1945, a short-lived position from which he nonetheless helped to craft the guiding legislation for science policy throughout most of the occupation period: Allied Control Law No. 25. Law No. 25 was taken seriously by OMGUS, with its strict ban on research of potential military use and reporting requirements for all basic research. Combined with the American authorities attempting—more than any other nation—to enforce a semi-thorough denazification policy, this was a significant incentive for scientists in the US zone to consider moving to other zones.

For those seeking work abroad, France and the United Kingdom were closer to home. The Soviet Union was ideologically daunting for many (though downright attractive to some) but offered relatively good pay at a time of near starvation amid Germany's shattered economy. Many would have liked to work with or even within the United States, clearly the world's new scientific superpower, but this initial discouragement of applied work was a real challenge. Of course, it is important to emphasize that no country had a unified message—during this same era, US agencies brought scientists to the United States in Operation Paperclip, among other programs, including some who were known Nazi enthusiasts. Still, official OMGUS policy was initially much less friendly toward researchers of industrial technology.

Those who aimed to rebuild German science took refuge in two rhetorical tactics: casting German science as "corrupted" or "perverted" during the Nazi period—and thus recoverable to a true science that might have better results—and emphasizing the distinction between "pure" and "applied" science. The first of these tactics, emphasizing "perversion" of German science under the Nazi regime, drew evidence direction from FIAT investigations that (quite to their surprise) came across as what they dubbed "scientific war crimes." On May 15, 1946, officials from the United States, the United Kingdom, and France—among them FIAT officials, representatives from the war crimes divisions of the United States and the United Kingdom, and professors from the Pasteur Institute and the University of Edinburgh—held a meeting to discuss what should be done with information gathered that "bore on the commission of war crimes by German scientists," in particular "inhuman experimentation on living men and women."[34] Instruction from the war crimes tribunals and legal divisions to FIAT eventually amounted to asking that all such evidence be forwarded on, and that they would deal with the issues, but for FIAT officials, the episode reinforced the concern that perhaps science itself had been corrupted by the Nazi regime.

One proposal created by the economic division of OMGUS in June 1945, titled "Technical and Scientific Research in Germany after the War," captures the ways in which this "pure" versus "applied" rhetoric allowed moderation between these control stances. "Pure" or "academic research, . . . defined as the expression of man's curiosity about the universe" unrelated to any "ulterior motive," is altogether for the good, the author of the proposal argues, and it is "the source from which all advances in technology must come." "Even the rarified fields of pure research" under the Nazi regime had been subject to the "perversion of German science," the "prostitution of science," however, and this was the justification for long-term scientific control (until "unmistakable evidence of a genuine change of heart in Germany"). As evidence of this "prostitution," the report's author cited German scientists collecting Yellow Fever samples to use as a weapon, "so causing a holocaust at a time best suited to themselves."[35] The contrast between what pure science could and should be, and what it had become because of the Nazis, was clear.

Meanwhile, American policymakers in Washington began to consider how rebuilding European science might be a way to knit together networks of elites who were friendly to the United States and therefore a way to build American "soft power" abroad. As John Krige has shown, this extended well beyond Germany and into every nation where Marshall funding might reach (i.e., most of Western Europe but not the eastern bloc, which was forbidden from taking Marshall plan funds in 1947).[36] These efforts often came through intermediaries, especially the Rockefeller Foundation, whose grants to the French Centre national de la recherche scientifique (CNRS) were explicitly aimed at reorienting the French scientific community toward the West, since its head (Frédéric Joliot-Curie) was a communist and might have preferred Soviet ties. On the whole, the American aim "to rehabilitate science in Europe was not only about providing material resources, but also about building structures and changing attitudes and values among scientists in line with democratic values."[37] In order to accomplish this, the United States funded scientist exchange programs, sponsored international conferences, and drew others into an international scientific community—"enroll[ing] national scientific elites on both sides of the Atlantic in the project of postwar European reconstruction."[38] These programs were positive, helpful, and often requested, adapted, and appreciated by Europeans. They were also explicitly a way to build American power abroad.

The Americans were not the only ones playing this game, however. The French and British also aimed to enroll scientific elites within Germany and the other Allied powers in ways that would support their own economic and scientific interests. These nations also sought to use science as a soft power to accomplish this goal. There are certainly important differences. France, for example, saw itself as working from a position of weakness to counter American strength by enlisting a Western European community under its leadership. Still, using science as a soft power was an international trend during this period.

French Efforts at Shaping Science through International Exchanges of Students

From the American and British perspectives, and in many histories of the occupation period, French policy in Germany seemed to be pure obstruction. Time and again, French representatives vetoed any attempts at overall policy and aggressive resistance of any centralized German institution.[39] This portrayal has more than a grain of truth to it but misses much of the complexity of French thinking on the question of handling German science. They, too, were occupied with how to accommodate and harness science for the purposes of the new French Republic. As happened in the American and British cases, French discussion about how to improve domestic science became entangled with planning for the future of German science.

There was no single funding entity for French science prior to the Second World War and little direct state support for science, a situation much like in the United States. Many ministries had some research mandate and their own programs, as did each of the branches of the military, but there was little coordination or centralized oversight. The CNRS was established in 1937 to serve in exactly this coordinating role.[40] The hope was that the CNRS could reverse an ongoing relative decline of French science measured (then and now) against Germany, the United Kingdom, and the United States. The exact timing and extent of this decline has been the subject of historical debate—some argue that France had relatively strong science prior to the First World War, despite weaknesses, while others argue it fell well behind before the twentieth century.[41] There seems to be general agreement that despite a renaissance in French science beginning around 1870, there was a significant decline in French science between the world wars.[42] This was especially true after the onset of the Great Depression, when funds from industry dried up. Rather than provide additional funding, the Ministry of Finance instituted a demand that research institutes pay the state for German laboratory equipment acquired through reparations from the First World War. The same demand related to reparations seizures after the Second World War would be equally unpopular.[43]

The mission of the CNRS was "to induce, coordinate, and encourage pure and applied scientific research pursued by the different public agencies or private enterprises, and especially to facilitate the research and scientific work of interest to national defense and the national economy."[44] Its bureaucratic position as a subsidiary of the education ministry did not afford it much power to accomplish these aims, however, as many departments outside education were reluctant to give up control of their own research units. Before any sort of meaningful compromises could be reached, Germany invaded, and the Vichy regime quickly reorganized the CNRS in March 1941 in a more limited role. As one example, they shut down the section for applied sciences and established an independent agency to promote science in overseas colonies. As I discuss in chapter 6, individuals within the CNRS worked to keep French scientists informed about Western scientific advances, but by and large the institution was neutered.

The early postwar years saw a return to bureaucratic struggles regarding the funding of French science. The CNRS faced new responsibilities, including a reaffirmation of its prewar mission of coordinating all French science and organizing French researchers in the French zone of occupation to investigate German science and technology, and handle scientific reparations. Meanwhile, the Ministry of Education, and thus the CNRS, had come under the control of the communist faction of early postwar France. Its new leaders, Frédéric Joliot-Curie and then Georges Teissier, aroused deep distrust in anticommunist factions inside France and abroad (including American agencies funding the reconstruction of European science by way of the Marshall Plan and other funding programs). Thus, in addition to diplomatic pressure from America to displace Joliot-Curie, other ministries—including the Commission for Atomic Energy (Commissariat à l'énergie atomique) and the Ministry of Technology—resisted the CNRS's efforts at truly integrating and coordinating French science.[45]

French scientists and policymakers were extremely sensitive about the ties between science and international prestige. In April 1945, the director of the École nationale supérieure des industries chimiques (a top university focusing on industrial chemistry) wrote to Joliot-Curie and Teissier in their roles as head of the CNRS, responding to a request for the names of students who could serve as *stagiaires* (trainees) in German laboratories, research institutes, and libraries (a plan discussed at length in chapter 3). Suitable students would be very rare, he wrote, and really only very experienced scientists and engineers should be sent—only they could see through the "inevitable subterfuge and deception" of the Germans.[46] The French should know, he continued, that, after all, they themselves had deceived occupying Germans over the preceding years, and all was fair to preserve "*le patrimoine du laboratoire, patrimoine matériel ou spirituel*."[47] Better not to send anyone than to send young students, because if they failed in any way—whether as spies or simply as scientists— this would hurt the prestige of France. The Économie nationale, writing in 1946, similarly emphasized the vital need to regain in reality "the preeminence that we claim."[48]

This combined crisis of confidence and sudden authority over a section of Germany led, quite naturally, to comparative questions: What had made German science more successful than French science? What could be learned? How could German science, now established as tremendously important for warfare, be controlled and prevented from contributing to resurgent German power? These questions were certainly not unique to France among the occupying powers, but the recent past and ongoing anxiety over the nation's role in the world gave them a particular urgency and importance.

Twin debates continued within the CNRS and other ministries, such as the Education nationale and the Control Commission for the French zone, about how best to reorganize French science, on the one hand, and how best to reorganize German science (either to promote, control, or cripple it), on the other. At a meeting to discuss reorganizing the CNRS in the immediate aftermath of the liberation of France, the chief criticism of French science was that it was too dispersed and that the CNRS of 1939 had inappropriately separated pure and applied science—France, they felt, no longer had enough researchers to support such divisions, whatever their desirability.[49] As they began analyzing German science in early 1945 in a series of internal reports on its nature, quality, strengths, and weaknesses, they felt that its key attribute was that it was centralized and efficient.[50] This was a lesson the CNRS could take to heart and use to promote a stronger role for itself in organizing French science.

Joliot-Curie, Teissier, and others began planning how the CNRS could best centralize French science. One proposal to draw reluctant, major research labs into CNRS control was to give these labs seats on the CNRS board. They faced many bureaucratic rivals, however, both within France and in their mission to the French zone of occupation. The mission of the Comité de coordination scientifique de la defense nationale (a.k.a. FIAT [France]), established on April 20, 1945, was not only to investigate German science and technology but also "to coordinate scientific studies undertaken by the military departments," founding liaisons between them and reducing redundancy.[51]

In the view of the École nationale, as expressed in a policy document in March 1946, science policy in Germany should pursue four ends:

1. To facilitate the missions of the occupation authority.
2. To permit the direction of German research by an international authority, or failing that, by a French organization.
3. To permit German researchers to survive out in the open, without any need to dissimulate.
4. To break the lines that exist between industry, commerce, and science in order to allow each to subsist to some degree without their combination allowing them to develop new weapons of penetration or attack against Germany's neighbors.[52]

This view was put forth in response to a proposal by the Greater Committee for Inventions (Comité supérieure des Inventions), who had proposed a clause in an eventual peace treaty barring all research. The École nationale, in contrast, was above all international-thinking—and that international future for France was very much as part of a unified Western European (as opposed to just Western) science. It recommended pushing for exchanging inventions and innovations through a professional, international organization; or failing that, a French one; or failing that, a "Franco-Belgian-Dutch-Luxembourgian" one.[53] In any event, French leadership was a priority.

This was another area where French policymakers felt they would have to go with practical, compromised solutions, rather than ideal ones. "From the point of view of security," argued the author of a report titled *The Organization of Scientific and Technical Research in Germany*, "it seems necessary to forbid or minimize scientific and technical research in Germany. . . . On the other hand, for those interested in the French economy, it would be preferable to maintain German research at an elevated level for the profit of France."[54] Chapter 3 discusses how this line of thinking led the French to pursue a policy of technical exploitation in place, but it also was relevant to broader science control policies. The French economy was felt to be too feeble to maintain scientific research on the same level as the Americans. Thus, the only path was to keep the Americans sharing their knowledge in the world market. This made supporting research in Germany indispensable, because as long as *someone* was producing research the Americans wanted, they had more incentive to stay engaged in European science. Promoting Franco-German/European research would also help develop an economic and scientific-technical collaboration with the United Kingdom, they hoped, and pull the United Kingdom away from its ever-closer ties to the United States.[55] A compromise between economic interest (pursued through science) and security, they felt, was both desirable and inevitable.

To this end, the CNRS proposed that within the French zone, "the Allies, and notably France, will have the right to introduce technicians of their choice in centers of German research, and impose exchanges of German students; they can thereby form in Germany French technical groups and thereby render truly effective control, and profit French institutions with the experience of German personnel."[56] With careful placement of the German technicians in French facilities, this could also generate positive impressions of French capabilities, improving international prestige.

The French military at least, and really most French institutions, were fairly paranoid about the possibility of German scientists using research for military purposes. They were not alone in this, though. The American planners in the economic subdivision of OMGUS shared French views about the need for long-term surveillance: "Year after year will go by without anything subversive in the scientific field being detected. Interest is bound to slack and it will be exceedingly difficult to maintain the necessary high standard of scientific control officers required. The only solution to this seems to be to grant research fellowships in German universities to suitably selected candidates on the understanding that they would be also observers for the Controlling Powers."[57] If Germans were not reformed and knit back into the world community, in a decade's time any practical surveillance would likely falter, just when it "should be strengthened and refreshed."

Contrary to the French reputation for vetoing all centralized German institutions, CNRS actually sought a centralized research board for Germany, funded by special taxes and proceeds from patents.[58] This organization, they felt, could then be intermittently inspected by interallied teams and might actually make such surveillance routine enough to survive politically. Further, this research would allow connections from German researchers to those in France and allied nations. There was some harsh disagreement within France about having this centralized research board. Education nationale pushed the opposite view in June 1946: "It is important above all else to avoid the centralization of German research; any general liaison between different research establishments is forbidden as dangerous, even if the work done by the establishment in question appears to present no military interest."[59] The CNRS countered that while there was some danger of research aiding German military aims, it would also help the French military and industry, and at the very least would create a hierarchy among German *savants* and technicians who could be monitored and perhaps co-opted in the event of war. This applied particularly to the Kaiser Wilhelm Institutes. Later in the year, when asked whether the CNRS was in favor of creating a society to promote German science analogous to the CNRS itself, a representative expressed harsh disapproval, insisting that the policy to follow was *"une politique de dispersion."*[60]

For CNRS officials, one imperative seemed to apply to both French and German science: that international connections must be reforged, immediately and extensively. Though no connections were to be allowed between German institutes, one of the chief duties the CNRS set for itself was to pair every German research institute with a French lab.[61] Further, as of July 1945, the CNRS favored "a modest and reasonable reconnection of German scientific relations with the rest of the world," leading to policies that included sending copies of the *Bulletin Analytique* (a bibliographic and abstracting service) to universities such as Freiburg, Tubingen, Mayence, Innsbrush, and Vienna, and engaging fully in the coordinated effort with the United States and United Kingdom to publish "FIAT Reviews" of German wartime science. The CNRS felt that German scientists had "too much neglected communication with allied *savants*" inside Austria—a primary goal was to connect Austrian and French scientists.[62]

This is particularly striking considering its contrast to the systematic exclusion of German scientists from the international scientific community following the First World War. International organizations expelled Germans during that war and disallowed reentry afterward. Journal delivery had been cut off in both directions during the Second World War and, as historian Michael Gordin has illustrated, contributions to English-language journals skyrocketed at the direct expense of the traditionally influential German journals, marking a tectonic shift in the international orientation of science.[63]

At home, French policymakers began looking around by March 1946 for partners, including both the stagiaires plan for placing French trainees throughout German scientific establishments and industrial research institutes, and thoughts of a Franco-British scientific alliance.[64] This became active policy in 1947 with negotiations for summer scientific student exchange programs with the Vacation Work Committee at Imperial College London, then the establishment of the Advisory Council on Science Policy to coordinate science exchanges.[65] Attempts to coordinate with the Swiss were more difficult because they had no centralized analogue to the CNRS. Negotiations with the Federation of American Scientists created opportunities for exchanges with the United States.[66] The Bureau Scientifique de New York was established in late 1947 and sponsored by the CNRS (though poorly funded, based on frequent requests for more support).

All of this collaboration created its own anxieties in a nation aiming to project the image of a modern scientific nation. A member of the Advisory Council on Science Policy worried that exchanges with the Federation of American Scientists might be "intended to carry off Europe's scientists."[67] When planning which students to send abroad, members of the CNRS's Comité de relations étranges (Committee for Foreign Relations) worried about sending students who were too young and impressionable, as "French identity is at stake."[68] Prestige was also on the line—the committee worried that France would lose face if foreign visitors were treated more poorly than French researchers abroad.[69]

There were many contradictions in French policy toward Germany. This was true in science no less than in any other field. Science would—if handled properly—grant prestige. It would knit together nations to counterbalance those with larger economies (i.e., America). It might even draw the British away from their wartime ties to the United States. To the extent that a unified "French" policy toward science can be said to exist, it was that expressed in a CNRS document around the end of 1945: the problem of creating serious limits on science is, above all, a problem of the control of the recruitment and training of young scientists.[70] By the late 1940s and early 1950s, key figures in each of the other Allied nations came to agree with this argument.

FIAT Reviews: Idealism Meets Intelligence

Symbolic of the importance of diplomacy and prestige in the exploitation of German science are the so-called *FIAT Reviews of German Science*. These reviews were attempts at a holistic account of wartime advances in a wide variety of academic science. German scientific leaders in Allied control were assigned to write them, and the final products were distributed throughout the world for free or at cost.

If this seems like an odd undertaking for military intelligence agencies, there were pragmatic reasons for creating these reports. The *FIAT Reviews* were born at least partly out of a need to find busywork for German scientists deemed too important to be allowed to roam free (potentially ending up in the employ of other powers, or, true to the paranoia of the time, perhaps even working on underground German war-related research) yet not valuable enough to be dealt with in a summary fashion. A group of scientists (40 percent of whom were reported to hold PhDs and speak English), for example, had been "frozen" in Heidenheim in July 1945 by the American military government and were still there in March 1946. They complained that their "scientific and technical talents [had] been completely wasted."[71] A top-secret report by the US Signal Corps to FIAT (US) discussing this problem emphasized that this was just one of many such groups throughout Germany. To find some use for these men, the report's authors suggested organizing them into a "FIAT Technical Institute" to prepare summaries of German scientific literature generated during the war, as well as English translations of more important articles. The reviews aimed to cover the entirety of the natural sciences, including twenty-seven volumes on medicine and pharmaceuticals, twenty-four on chemistry, sixteen on physics, eight on earth sciences, seven on mathematics, and four on biology, to be printed in translation abroad and as *Naturforschung und Medizin in Deutschland, 1939–1946* inside Germany.

For British policymakers, the *FIAT Reviews* were useful as a semi-philanthropic front for any bad press that BIOS might generate. Authoring these reviews could "hardly be looked upon as exploitation as it is done with the very willing co-operation of the German authors who are anxious to see their work recognized in print."[72] It was not only the domestic press and international observers who might need to be placated. By 1947, the CCG/BE was taking a firmer stance against exploitation, to the extent that by January 1947, "whereas almost anything was at one time considered fair game, investigators [were] now discouraged from enquiring into details of new peaceful German inventions."[73] The *FIAT Reviews* were an easier sell, even within branches of the British government.

German scientists were eager to author these reviews, in part to reconnect to the world's scientific community, in part to receive a paycheck in a brutal economy, and in part to rewrite their own collaboration with the Nazi government. Future Nobel laureate in Physics Walther Bothe and Professor Siegfried Flügge were asked to inventory nuclear physics, their area of expertise, in 1947. The *FIAT Reviews*'s phrasing balances carefully between demonstrating the scientists' value and ignoring any of the context in which the research took place. Historian Mark Walker has dubbed the *FIAT Reviews of Nuclear Physics* "apolitical apologia written for scientists by scientists," in that they carefully avoid the military motivations and applications for German research.[74] In all, the more than four hundred pages included contributions from almost every scientist involved in wartime German nuclear research. The American authorities were certainly nervous about the publication, and Ralph Osborne, chief of FIAT (US), sent the US War Department staff an early draft and warning about the limits of his control: "If many or major changes are required a difficult diplomatic situation may be anticipated. Actually, the only manner in which any control is exerted over this manuscript is through the present publisher being located in the US Zone."[75] Whatever their reservations, the report was both published and distributed along with the FIAT reports.

Curiously, even more controversial than nuclear science were the "humane sciences." In December 1946, General Gaston de Verbigier de St. Paul, director of FIAT (France), wrote to Colonel Osborne requesting cooperation in researching and writing a "FIAT Review of Humane Sciences."[76] Osborne's primary objection to such a review was expediency—he felt "a very strong sense of the urgency of time" in getting the *FIAT Reviews* series published and wished to avoid expanding its scope at that relatively late date. Beyond this, however, was "the controversial nature of any findings in these sciences" that made it "not possible for me to cooperate with you. . . . but [I] would be happy to see the [CNRS] undertake this as a French effort."[77] The exact definition of "humane sciences" under consideration and the reasons for its controversy are not entirely clear. In any case, the reasoning speaks directly to the rationale of the FIAT Reviews. These were not to be controversial reports, or to cover controversial material. They were to inform, certainly, but only within the limits of apolitical science.

Despite paper shortages and difficulties finding local printing facilities, FIAT (US) produced 1,100 sets of FIAT Reviews to be shared equally among the three FIAT organizations. Each received 250 copies, while thirty-three more were presented as gifts to UNESCO (United Nations Educational, Scientific and Cultural Organization) for other nations to utilize, fifty-seven copies were earmarked for each occupation zone, and twenty additional copies were reserved for each zone's military government. Aside from the cost of such an undertaking, the equality with which France was treated, the designation of many copies for use in improving German science, and the continued exclusion of the Soviet Union mark the extent to which the American government remained aware of the diplomatic role these FIAT Reviews could play. As is discussed in the next chapter, the Allied powers struggled with the immense scale of industrial research acquired by FIAT and BIOS efforts, but in the domain of the natural sciences at least, distributing information was worth substantial costs just to be seen doing it.

Conclusion

The legacy of the Nazi regime on German science stretched far beyond the occupation period. Scientists were deeply involved in building up both West Germany and East Germany, and in the context of the early Cold War, many questionable (or downright abhorrent) histories of aiding the Nazis were swept under the rug. For decades afterward, German nuclear scientists promulgated the myth that they had somehow undermined the Nazi nuclear program, and that was why Hitler never got atomic weapons.[78] As I examine in the next chapter, the rush to gain prestige and influence by rebuilding the infrastructure for scientific communication (journals, abstracts services, etc.) was similarly marked by pragmatism and compromise over idealized denazification.

Each of these nations was eventually willing to give up on purging Nazis from government—and even to take former Nazis among the scientists imported to their own homelands—in part because of the near impossibility of doling out blame at all fairly and in part because science was now seen as too important to let fall behind. Policy documents in each nation discussed scientific manpower as a national security resource in terms similar to uranium deposits or air bases: a zero-sum resource, where your gain is your enemy's loss. Allowing Germany to rebuild military power was initially unthinkable, but having access to new pools of talent to hire away was extremely tempting. This was particularly true for the three Allied nations most damaged by the war, but America also sought to control what scientists it did not need for its own universities and industrial research centers.

Thinking about the role of science in society was more complex than seeing it as simple ammunition for the Cold War, though. Throughout the early twentieth century, organizations of scientists consciously promoted science as an international good, full of cooperation, meritocratic and democratic governance, and open communication. As scientists gained more political power as a result of the war, they employed this rhetoric in battles at home and in Germany about the importance of funding "pure" science for the health of a democratic, open society. In this way, too, developments at home mirrored and shaped those in rebuilding (and co-opting) German science.

If the First World War taught the Allies that severe monetary reparations were a bad idea (and that perhaps scientific and industrial "intellectual reparations" might be a better path), they likewise were taught that cutting Germans out of international science was not a great plan. In sharp contrast to the continued isolation of German science after the First World War, in the late 1940s each of the occupying powers instituted programs designed to rebuild German science in their own image, connected through both formal and informal networks to their own labs and universities. This took different forms in each nation and occurred at different times, and in each nation factions resisted these programs. Still, science became a potent tool of diplomacy in this early postwar period.

Providing funding for exchanges, organizing international scientific conferences, sponsoring research, and building networks of benevolent scientific elites allowed some measure of leadership (and thus control). The United States had the most resources available for this form of soft power diplomacy, with its powerful economy and scientific infrastructure advanced by the war, and through official governmental policy and backdoor liaison with independent groups such as the Rockefeller Foundation, it used these tactics to knit together a Western scientific world oriented toward America.[79] It was not alone in doing so, however. For the British and French, too, the control of science—at home and abroad—was fundamentally about the formation of the next generation of scientists. This applied to French thinking on how to keep German scientists from helping resuscitate a feared Teutonic military aggression, but it also applied at home, where the reestablished CNRS fought bureaucratic battles with those who resisted centralized coordination of the nation's scientific future under the control of communists. The lack of resources during rebuilding was a serious concern, and French policymakers worried that an inability to fully reciprocate in international exchanges with full hospitality would negatively affect one of the primary goals of this science diplomacy: prestige and appearance of leadership. It was still deemed worth pursuing, however, and even worth pushing Germany to reconnect to this community (whether internationally, bilaterally, or as part of a Western European community to balance against the Anglo-American and Soviet poles). Soviet hegemony over East Germany (and its sphere of influence generally) was more direct, but there, too, science had a unique prestige, since communism itself was supposedly a "scientific" method of government.

What sense is there, then, in having the same institutions handle both exploitation of German technology and the reconstruction of German science? In part, it was a realization that science and technology are fundamentally linked in complex ways, and that distinctions between science and technology are to some degree a game of semantics. In part, it was a shortage of scientifically trained bureaucrats willing to work in occupied Germany, and any technical knowledge was deemed better than none. It was also a conscious effort by these scientist-bureaucrats to carve out their own sphere of society (both at home, for example, in promoting a National Science Foundation run by scientists, and abroad), even when they recognized that the intelligence services might actually be the most logical group for the job. Finally, the image of neutral, apolitical scientists was useful—both for the German scientists themselves and for the Allied powers—in promoting FIAT and its ilk as semi-philanthropic organizations working to the benefit of all.

6

Documentation and Information Technology

Dealing with Information Overload

In a real sense, for at least most American and British planners, the exploitation of German science was an information problem. Germany had lots of valuable information they wanted, so they sent teams to capture it using cutting-edge information technology. This information would become available to all by way of both government-backed and private publishers, who could use these same breakthrough technologies to reduce the price of copying and sharing so much that almost anyone could afford them. Combined with the latest bibliographic tools and methods, this information would almost become—in multiple senses—free for all humankind. New industries, technologies, and scientific breakthroughs would surely result, simply by making this valuable treasure trove available to all.

Technological enthusiasm like this—the idea that information technology would seamlessly turn bulk data into useful knowledge—has a long history and very much lives on today. So, too, does the general problem these information technologies promise to solve: information overload. Since at least the late nineteenth century, exponential growth in publications about science and technology has overwhelmed those who have sought to keep up with advances in even focused subfields. By the Second World War, an American physicist born in 1900 would have witnessed enormous changes in the world of science. The year he was born, all American universities combined awarded fewer than three hundred natural science PhDs.[1] That number had increased 40 percent by 1914. Between 1920 and 1940, it shot up another 600 percent. This exponential growth in global science can be seen in any number of metrics (the number of scientists, science journals, journal articles, etc.).[2]

During the first half of the twentieth century, networks of librarians, scientists, and entrepreneurs (many of whom we might today consider early "information scientists") developed tools for tackling this problem of ever-expanding science. This movement, dubbed the "documentation movement" (and, in the United States, sometimes called the "special libraries movement"), sought to accomplish nothing less than cataloguing all human knowledge in a quickly accessible format.[3] These utopian, Enlightenment-esque visions were to be accomplished in part by figuring out how to index, categorize, and sort data; build bibliographies; and set standards for metadata.[4] At information scientists' disposal were cutting-edge information technologies with enormous potential, including microfilm, card catalogs, typewriters, and even early forms of mechanical computers.

In order to understand why British, American, and to a lesser degree French and Soviet planners saw the exploitation of Germany through this somewhat naive, technophilic lens, you have to understand this context of documentation, scientific internationalism, information technology, and government involvement in spreading science. As Allied leaders sought information about German technology during the war, and in planning the exploitation programs afterward, they used the technologies available to them and hired the personnel best equipped for the job. This meant librarians and documentation enthusiasts, who eagerly entered government service, bringing their ideas, assumptions, and ambitions with them. If wartime hostilities dammed the international flow of scientific information to a trickle, these documentarians knew full well that postwar investigations would bring a flood. Yet with the help of these new technologies, and now the full support of governments they had long sought to involve in their crusade, they set to the task.

In this chapter, I explore these roots of the investigations in Germany and the consequences for scientific information systems during the Cold War. Exploitation programs played an important role in pushing governments to grapple with the circulation of technical information. With science now a national security priority, the United States and Soviet Union, in particular, invested massive funding into research and development on scientific bibliography, data processing, and early computer technologies. In so doing, they laid the groundwork for what would become known as library and information sciences. Meanwhile, intelligence agencies focused on science and technology throughout the Cold War, and both benefited from and sponsored information technologies.

Finally, this chapter ties in the private sector entrepreneurs who saw enormous profit potential in this ever-growing world of scientific communication. Between government investment and expanding scientific societies, there was a lot of money in science publishing. Translating and reproducing reports about German technology posed a prime opportunity for a few firms—and as the scientific translation industry took off, it paved the way for English to take root as the international language of science.

The Documentation Movement during the Second World War

As scientific research multiplied exponentially in the late nineteenth and early twentieth centuries, librarians, scientists, and industrial groups started to plot how to control the flood of information. The founding of the International Federation for Information and Documentation (FID) in 1895 in Belgium by lawyer Paul Otlet was a landmark for this movement, which came to be called the "documentation movement."[5] Otlet found an eager audience in libraries across several nations, and not just in universities. Private industrial research laboratories, especially in the United States, increasingly invested in their own special libraries. By 1910, articles in *Library Journal* and industrial trade journals discussed the need for the library as an adjunct to the industrial laboratory.[6]

The FID organized five international conferences on bibliography in 1895, 1897, 1900, 1908, and 1910. Initially, planners dreamt big: an international bibliography of all science, organized by a grand Institute of Bibliography. This institute folded after only a few years, however. Still, during this time, the FID drove advances in bibliography, database management, and information retrieval. As the FID enlisted institutions including the Library of Congress and the British Museum in its mission, allied and competing groups sprang up around the globe. While their exact purpose and definition of the term "documentation" varied slightly from place to place, the International Institute of Documentation's definition suggests the ongoing raw ambition of those involved: "the assembling, classification and distribution of documents of all kinds in all areas of human activity."[7]

The movement was far broader than just Otlet and the FID. American scientist-turned-documentalist Herbert Field founded another ambitious center for universal scientific bibliography, the Concilium, in the 1890s, hoping to establish international standards and eventually cover all science.[8] The British Royal Academy considered expanding its bibliography efforts during the same period. German science publishers, including those working on collections of abstracts, grew powerful in serving both domestic audiences and for alerting the world to German scientists' advances. The American Chemical Society began the influential *Chemical Abstracts* for its membership in 1907. Not all of these attempts at organizing science had equally internationalist aims, as more traditionally powerful nations (e.g., Britain and Germany) sought to shore up their prestige and influence and some (e.g., *Chemical Abstracts*) focused on specific fields rather than on all science. Still, the documentation movement was international in scope and often in aims, and demonstrates how widespread the feeling was that the scientists were facing a crisis of information overload.

The First World War

The First World War was both a disaster and a crucial step for the documentation movement. On the one hand, the war shattered the world of international science. As the wartime alliances developed, the British and French scientific communities expelled Germans from international scientific institutions they had previously dominated, and anti-German sentiment outlived the war itself.[9] Embargoes and submarine warfare almost entirely stopped the flow of journals and books between Axis and Allied blocs, and wartime conditions shuttered the Concilium and other bibliographic enterprises.

On the other hand, the war made science a priority for governments, opening the door for meaningful state support. Historians have dubbed the First World War "the Chemists' War" because of the importance of poison gas, explosives, and other scientific contributions. State support for science was extremely limited before the war, especially in the United States, and enterprises such as the Concilium depended on philanthropy and end-user subscriptions (and still accumulated large debts and eventually failed). The First World War began a process of convincing governments that funding science was a national security imperative. That process would slow in the interwar years, as the Great Depression led to austerity policies, but set the stage for dramatic expansion after the Second World War. Specifically in the realm of scientific information systems, the First World War also had the immediate effect of bringing heavy government investment in sharing science *among* allied nations, even as it broke ties across battle lines.

Among the Western powers, sudden isolation from Germany's academic and industrial research centers was a rude shock and provoked both internal and cooperative responses. In the United Kingdom, this took the form of the Department of Scientific and Industrial Research, an entirely new department of the government founded at the suggestion of the Board of Trade and

Board of Education "to finance worthy research proposals, to award research fellowships and studentships [in universities], and to encourage the development of research associations in private industry and research facilities in university science departments."[10] In June 1916, President Woodrow Wilson requested that the National Academy of Sciences set up a group to coordinate and apply scientific research to national defense, and the result was the National Research Council (NRC). France did less institutional reorganization, with so many resources (including young scientists drafted as soldiers) being sent to the front lines, but did spare some resources for existing laboratories at institutions such as the École supérieure de physique et chimie industrielles de Paris. Russia, at least after March 1917, was busy with internal revolution and questions of whether scientists should be purged as bourgeois intellectuals.[11]

On the diplomatic front, the need for scientific information was enough to overcome British suspicion of the French. Similar to the French needing to convince the British to coordinate efforts on FIAT after the Second World War, it was the French who instigated the exchange of science attachés during the First World War.[12] In December 1915, French Minister of Public Instruction Paul Painlevé sent a formal request to Britain's David Lloyd George for a formal exchange of scientists. The British Admiralty dragged its feet on this request for months. Lloyd George agreed to collaborate on weapons research and development in February 1916, but it was not until September that the British Foreign Office overruled the Admiralty's hesitations. In December 1916, Louis de Broglie, a French physicist who made fundamental contributions to quantum theory, arrived in London. Commander Cyprian Bridge, a naval officer (and later admiral), represented the Admiralty's research wing in Paris.

Despite the slow start, British and French cooperation was very productive. British officers who worked with French colleagues admitted that the French were very courteous and helpful. Together, the French scientist Paul Langevin and the British physicist Robert Boyle played crucial roles in developing sonar. Sonar, in turn, was vital in fighting submarines that were devastating Allied morale and seriously hampering shipping across the Atlantic.

When the United States joined the war in April 1917, research-sharing agreements with Britain and France were an early priority. The NRC organized the Research Information Committee in December 1917, which was tasked with "cooperating with the offices of Military and Naval Intelligence in securing, classifying and disseminating scientific, technical and industrial research information, especially relating to war problems, and the interchange of such information between the Allies in Europe and the United States."[13] The basic goal was to reduce the hassle of sending teams to Britain and France every time a question arose regarding the four main topics in which America needed to catch up: submarine detection, chemical warfare, trench warfare, and aeronautics.

America's first permanent scientific attachés, William Frederick Durand and Henry Andrews Bumstead (assigned to Paris and London, respectively), organized teams of American researchers. These teams braved submarine-infested Atlantic waters in order to meet with their counterparts on British and French research teams. British officials noted privately that despite plans for a mutual exchange of information, far more American researchers than written reports of results were making it to London.[14] Still, if this kept the Americans in the war, it was well worth the cost. Reports from these scientific attachés began streaming back into the United States in the weekly diplomatic bag. These reports were passed on to the Research Information Committee, which was tasked with indexing the contents, making copies, and distributing these copies to US military agencies.

There is reason to be skeptical about what exactly these reports accomplished. Getting information to people who could use it without also overwhelming those people with lots of less useful information was already such an intractable problem before the war that it had launched the documentation movement. There was also the obstacle of how much could not be captured in written reports, a problem later seen in FIAT and BIOS. Also similar to FIAT, testimonials at the time vary sharply as to the value of these interallied communiqués. Bumstead, the science attaché in London, effusively praised British cooperation and felt he sent back *potentially* invaluable reports but worried whether assurance from Washington that they were valuable "wasn't all camouflage" for them having little impact.[15] Robert Millikan, the famous physicist and key player in the NRC, reported reading the reports and gaining useful information from them. Still, as historian Walter MacLeod put it, "methods for capitalizing upon scientific intelligence were even less sophisticated than the channels for transmitting it. The immediate consequences for the war may have been limited."[16]

Interwar Period: Microfilm Mania

As the war ended, the documentation movement set back to work, balancing ambitions of unifying and coordinating international knowledge with fierce nationalism engendered by the war and the new expectations governments imposed in exchange for expanded investment. Otlet, Field, and others successfully campaigned to make scientific information a priority for the new League of Nations, though the US Senate's rejection of the league undercut the reach of the ensuing programs. Field sought to restart his Concilium, this time with funding from two of the most important sources for science funding prior to the Second World War: the Rockefeller and Carnegie Foundations. In the interwar years, private firms also invested heavily into new libraries and information centers that hired documentalists and special librarians to turn the new bibliographic techniques and library practices toward industrial research. DuPont's Intelligence Department, founded in 1919, is just one example among many.

The grand ambitions of the documentation advocates during the interwar years were fed by enthusiasm for the seemingly limitless possibilities of a new (or at least, newly practical) technology: microfilm. Though microfilming technology had been around since the mid-nineteenth century, and had been promoted as a solution to scientific record-keeping by astronomer James Glaisher in 1851, the technology did not achieve widespread popularity until the late 1920s, when Eastman Kodak developed the economical Recordak microfilm camera.[17] Initially, Kodak's primary market was banks, who sought a way to photograph and store checks. In the 1930s, the technology spread to libraries, including the US Department of Agriculture Library, the Library of Congress, the British Library, and the Bibliothèque nationale in Paris.

For those invested in information dispersion and retrieval, microfilm offered a relatively cheap way to store, copy, and transport large amounts of information, leading to visions of libraries transformed to rooms of viewing machines with a central card catalog, the microfilm for each entry simply taped to its card.[18] At the International Exposition and the World Congress of Universal Documentation held in Paris in August 1937, the Rockefeller Foundation, American Chemical Society, and University of Chicago together hosted an exhibit promoting microfilm as a solution to the growing information problem. Similar conferences drew prominent authors, politicians, scientists, libraries, and others interested in information management, all of whom were captivated by the promise of microfilm.[19]

Interwar Britain

In the United Kingdom, for example, scientists including the Oxford chemist Henry Tizard—later a noted (and knighted) scientific attaché and diplomat—

worked with the Department of Scientific and Industrial Research (DSIR) and government personnel to ensure that British research would never repeat the wartime bottlenecks in scientific knowledge. One important institution forged to this end was the Association of Special Libraries and Information Bureaux, or Aslib.[20] In the words of historian Pamela Richards, "the history of documentation—or information science, as it is commonly called today— is in England largely synonymous with that of Aslib, its members and their activities."[21] Other centers existed, including the British Society for International Bibliography (founded in 1927 and allied with Otlet's International Institute of Bibliography), but Aslib was the locus of activity.

During the interwar period, Aslib's activities included taking a census of what specialized libraries existed in the country, synchronizing standards across them for bibliography and metadata, and plotting out ways to duplicate and distribute foreign journals more efficiently. The Science Museum Library in London was a major beneficiary and leader in this charge, as it became a central location for coordinating the duplication and distribution of scientific journals.[22] As the rumblings of the Second World War approached, Aslib's planning expanded to include which bunkers and rural estates might house libraries during bombing raids as well as a renewed effort to translate, abstract, and distribute Russian and Japanese scientific publications.

A key figure in British documentation was John Desmond Bernal, a Cambridge scientist and pioneer of X-ray crystallography. Bernal was an ardent leftist and visited the Soviet Union several times as a member of the Communist Party of Britain. Tied to these political commitments, Bernal sought to reform science journal publication, argue for planning science around improving the lives of the working public, and build "a system in which all relevant information would be available to each research worker in an amplitude proportional to its degree of relevance."[23] Bernal's vision never came to pass, but he represents a few trends among the documentalists: international character, as he worked with both American and Soviet bibliographers, and leftist politics, common among many (though certainly not all) of those applying scientific planning to science publishing.

Interwar America

In the United States, Watson Davis founded an American Documentation Institute (ADI) in 1937, and microfilm stood at the heart of much of its planning. Davis was formerly the editor of Science Service, an organization dedicated to science education and popularization through means of developing journalists who understood and could better write about science. The ADI drew funding from the Carnegie, Rockefeller, and Chemical Foundations, and coordinated with the Library of Congress, Departments of Agriculture and Education, and various public health agencies.[24]

One source of interwar-era funding for experiments in microfilm and documentation was the original American effort to seize German science during wartime: the Chemical Foundation. As mentioned in the introduction, during the First World War, the United States government seized all patents and copyrights owned by German nationals, placing them into temporary custody of an Office of Alien Property Custodian. The Alien Property Custodian was originally mandated to safeguard and manage this German property with an eye toward returning them to their original owners after the war, but as anti-German sentiment grew throughout the war, this purpose faded. In a bid to free the American chemical industry from reliance on the major German cartels, US firms came together to form an organization called the Chemical Foundation, which bought all of these German chemical patents from the Alien Property Custodian. The Chemical Foundation then licensed these patents to American firms very cheaply, bringing in a moderate income used mostly to fund the foundation's operations.

That same income stream was an important source of funding for the documentation movement in the mid-1930s, in the form of donations to the ADI. Francis Garvan, head of the Chemical Foundation, believed in the cause to such an extent that he personally donated 15,000 USD in 1935 in addition to the foundation's 100,000 USD in funding.[25] Watson Davis, as head of the ADI, used this funding to buy and develop new microphotographic equipment, and work on integrating microfilm-based indexing and storage into the work of Science Service.

Outside of the ADI, the National Academy of Sciences was one of many other groups trying to build solutions to the growing logistical problems of importing and even translating foreign science. In 1937, a committee from the National Academy of Sciences that included Vannevar Bush, Ludvig Hektoen (chairman of the NRC), and Irvin Stewart (former chairman of the Federal Communications Commission) investigated whether it would be useful to set up a service to microfilm and distribute any copyright-free foreign scientific publications.[26] They found that despite individual scientists insisting that such a service would be crucial, actual demand was never as much as they hoped. One key hurdle would be familiar to anyone who has ever worked extensively with microfilm: the film was not always clear, the equipment was slow and uncomfortable to use, and reading through long reels was altogether unpleasant.

Since at least the late nineteenth century, American science had been growing in both scale and prestige relative to traditional centers in Britain, France, and Germany. In the interwar period, some influential scientists such as Charles Davenport began using their influence with funding agencies to push documentation efforts (among other aspects of science infrastructure) to become American services rather than internationalist services funded in part by Americans.[27] A key example is Herbert Field's Concilium. As Field tried to rebuild his bibliographic/abstracting service in the interwar years, he became torn between those, like Davenport, with a more nationalistic approach and those who sought to rebuild international science (e.g., by building on the wartime International Research Council that coordinated with Britain and France).

This tension built throughout this period without any clear resolution, with American "Big Science" growing more prestigious and American individuals and agencies seeking to exert more influence on the world stage in science and through science. This trend would come to a head in rebuilding Europe after the Second World War, where historians such as John Krige have brought attention to explicit efforts by the US government and the Ford and Rockefeller Foundations to reconstruct European science in an American image.[28] This impulse existed in rebuilding after the First World War as well—it was the scale that differed.

Interwar France

France remained the home of many of the internationally oriented documentation efforts during the interwar years, and much of their interwar programming centered on promoting microfilm as a space-, cost-, and time-saving technology. It is no coincidence that the International Exposition and the World Congress of Universal Documentation, where microfilm was so eagerly evangelized to scientific societies and libraries, was held in Paris in August 1937.

Jean Gérard, a chemical engineer by training, turned the connections he built up as a wartime chemist in poison gas research into a major organizing role in international science, including documentation. He was a founding member of the Société de chimie industrielle (Society for Industrial Chemistry) in 1917, which sought to develop a comprehensive chemical library and an information service for academic and industrial chemists, and to publish journals in the field. The members of the Société de chimie industrielle, with Gérard's active involvement, helped found the International Union of Pure and Applied Chemistry in 1919, an interallied group that was part of a broader reorientation of international science during the war to exclude Germany. This same group, now under Gérard's leadership as secretary-general, played an important role in helping German scientists reintegrate into international chemistry by inviting Fritz Haber to lead a delegation to their 1930 conference. Gérard, a strong believer in international cooperation in science, campaigned actively through his now-extensive network of international chemists to ensure that the Germans received a polite reception.

Documentation was part and parcel of Gérard's campaign to knit together the world's chemical research communities. When the League of Nations' International Committee on Intellectual Cooperation proposed the establishment of an International Office of Chemistry to coordinate standards for bibliography within chemistry, Gérard emerged as a key figure. In the late 1920s, Gérard helped organize the construction of the Maison de la Chimie in Paris, an institution that would include a center for documentation and a major chemical library, and would serve as an organizing force for conferences around the world. In the 1930s, with the Maison de la Chimie finished and Gérard in place as its administrator, he founded the Société de productions documentaires (SOPRODOC) in its basement, which served as an indexing and abstracting service for chemical science, much like that orchestrated by Aslib in the United Kingdom.

In 1937, Gérard helped organize the World Documentation Congress in Paris to advance the field. This meeting brought in a wide range of guests from around the world, ranging from prominent scientists such as John Desmond Bernal to authors such as H. G. Wells, who spoke on how documentation efforts were key to making available the "world brain."[29] Throughout the interwar period, Paris was an international hub for cataloguing and distributing scientific knowledge, and Jean Gérard was at its core. The Second World War, however, would soon tear Gérard's work to shreds.

Interwar Soviet Union

The Soviet Union, born out of the First World War, had enormous tasks to accomplish in the 1920s and 1930s—nothing less than the radical transformation of an entire society. Even with these existential questions facing Vladimir Lenin and other early Soviet leaders, a desire for well-organized access to international science, heavily influenced by documentation principles, was a priority. Lenin himself had studied Otlet's Mundaneum, the British Royal Society's International Catalogue of Scientific Literature, and other documentalists' work during his years in exile before the war. Seeing control of information as crucial for the new state, Lenin began planning an integrated library system across the USSR and invested in bringing the most cutting-edge information technologies (e.g., typewriters, filing systems, Linotype machines, microfilm systems) into the new bureaucracy, buying the newest American and European models.[30]

Seeking information on international science, the Soviet Union established the Bureau of Foreign Science and Technology in 1921, focusing primarily on gathering information from Germany and sending it into the Soviet library system. Lenin himself proposed standards for abstracting, indexing, and acquisition policies, drawing upon his familiarity with the documentation movement. German scientists, for their part, were still reeling from Western boycotts of German scientific institutions and eagerly embraced information exchange with the Soviets. A committee that included Albert Einstein, Max Planck, and Adolf von Harnack, the head of the Kaiser Wilhelm Society, was formed in 1923 specifically to promote scientific exchange with the Soviet Union. Both Soviet and German groups, public and private, translated German works into Russian (and some from Russian into German). A bureau within the Soviet People's Commissariat for Education, KOMINOLIT, subscribed to foreign scientific and technical journals, even seeking to find back issues from 1914 onward. After the 1922 Treaty of Rapallo, German military and industrial specialists (including those from major institutions such as Krupp and Juncker) set up facilities in Soviet territory, helping Germany avoid limits imposed by the Treaty of Versailles and advancing Soviet mastery of these technologies.[31]

Despite these significant efforts to build a Soviet information economy with many parallels to the documentation movement in the West, these interwar Soviet efforts faced serious problems by the late 1920s and into the 1930s. The place of scientific elites and prewar institutions such as the National Academy of Sciences was continually up for negotiation in the new political regime. Radical student groups, Marxist scholars, and party leaders questioned their loyalty, and saw ties to Western scientists as potentially corrupting, or at least evidence of bourgeois sympathies. As Stalin began asserting control and purging any potential sources of autonomous authority, connections to foreign science became increasingly dangerous and ideological purity became paramount.[32]

Still, going into the Second World War, the Soviet Union had at least some infrastructure in place to acquire, sort, copy, and disseminate scientific and technical information to pressing problems. The recent history of learning from German science and technology, building on a much longer history of at least occasional Russo-German scientific engagement, undoubtedly also influenced Osoaviakhim and other exploitation efforts in the early postwar years.

The Second World War and the Movement of Scientific Knowledge

The start of the Second World War once again split the international scientific community that had only begun to reintegrate. The rise of the Nazis led to an early exodus of Jewish scientists in the 1930s, much to the benefit of the Western nations. By the late 1930s and early 1940s, few scientists moved between Axis and Allied nations. Scientific journals, letters, conference proceedings, and books from the opposing nations became scarce right when they were most valuable.

This was especially true after Germany invaded and occupied much of Western Europe in 1940, including France's research institutions. Among the many science and technology collections the Nazis seized was Paul Otlet's life's work, located in his Mundaneum in Paris. They seized some of his books, then destroyed sixty-three tons of books, periodicals, and the card catalogs and organizational tools he had spent fifty years developing. By April 1941, British officials reported that no French, German, or Italian periodicals had entered the country since May 1940, and only infrequent deliveries by the Air Ministry provided Soviet science.[33]

Each nation's responses to scientific information crises of the First World War, then, were quickly put to the most extreme test. The documentation enthusiasts' developments—both institutional (e.g., Aslib) and technological (e.g., microfilm processing)—would be crucial for managing the flow of wartime information within countries, among allies, and even over enemy lines. Science journals and technical papers were far from the only information in need of processing, of course. As historian Jon Agar has argued, "the conflict was also an 'information' war. . . . [The war] presented massive problems of managing extended technological systems. People and materiel had to be organized at immense distances."[34] Government agencies sought to coordinate their entire economies, equip and mobilize armies, combine scientific research capabilities with allied nations in conditions of strictest secrecy, and discover everything possible about opposing and neutral nations. This led to a massive investment in both office machinery and bureaucracy, and these agencies drew upon whatever personnel they could find with documentation / information science / librarianship skills at their disposal.[35]

With the efforts to take German science and technology, all of these trends came together: bulk information management, newly assertive governments, scientific communication, the increasing importance of science for warfare, intelligence agencies, and diplomacy through science. These FIAT-related programs would prove both subject of and amplifier to documentation planners' successes and failures to smooth out bottlenecks in dealing with massive volumes of scientific and technical information.

The United Kingdom Enters a Scientific War

In the United Kingdom, the DSIR, the British Library, and Aslib worked together to make do with the information they still received from continental Europe. Networks of spies and diplomats in neutral countries could still buy copies of German journals and books, then ship them to the United Kingdom in diplomatic pouches. Microfilm was again a key technology for allowing these diplomatic pouches to transfer large amounts of information. Still, this was relatively small scale, and these pouches could never replace the thriving trade in journals that had fed the world of international science before the war. Aslib turned to two options: better organizing what information was available and creating microfilm duplicates to distribute throughout the United Kingdom and the United States.

As a result of their interwar census efforts, Aslib and DSIR were collectively aware of the nation's specialized libraries, so as war preparations got under way in earnest, they were able to craft plans to reduce redundancies among them. By June 1940, Aslib began investing resources into producing "Wartime Guides" to scientific information, listing where in the United Kingdom each available journal and book could be found, by subject.[36] Still, knowing where to find information is of little help if the documents were already in use elsewhere. Keeping up with orders was a major challenge as military research ramped up into new fields—radar, aeronautics, and eventually nuclear weapons research prominent among them. The Science Museum alone received more than double its prewar requests for borrowing from its collections in physics, chemistry, and engineering.[37]

Photocopying the necessary documents might seem like an obvious solution, but a keen respect for intellectual property law stymied Aslib's plans to pursue it. The Patents, Designs, Copyright, and Trade Marks (Emergency) Act of 1939 authorized copying imported copyright materials only in cases that could be considered urgent, and the Patent Office interpreted that very literally.[38] In October 1941, more than two years after Great Britain declared war on France, Aslib had yet to overcome the Patent Office's objections, despite repeated secret memos to the War Cabinet's Scientific Advisory Committee. Finally, in December 1941, a meeting hosted by the DSIR between the government libraries, Her Majesty's Stationery Office, and Aslib came to a decision to reproduce enemy periodicals en masse via microfilm (at least until shortages of microfilm forced other methods). The small Aslib Microfilm Service (AMS) set up shop above the Victoria and Albert Museum with five American Kodak Microfilm cameras, the best then available.[39]

American Parallels and Anglo-American Cooperation

American institutional responses to the need for German science were similar to those of British programs. Library associations formed the Joint Committee on Importations in 1939 to centralize bulk orders of journals from the continent, hoping that at least some would make it through, but these were less and less successful. As a result, American agencies in need of German scientific materials turned toward a newly available resource: a newly formed intelligence agency. The Office of Strategic Services (OSS) was a wartime, civilian intelligence agency founded in June 1942 (six months after the Pearl Harbor attack brought America officially into the war) and a predecessor to the Central Intelligence Agency.

The OSS quickly set up its own unit, dubbed the Interdepartmental Committee for the Acquisition of Foreign Publications, which would be responsible for coordinating the purchase of journals through agents in neutral countries.[40] This program was technically accessible to any American library through the Library of Congress, though it was not widely advertised. The OSS also targeted "open-source intelligence" (meaning publicly available; e.g., newspapers or government publications) about both enemy and allied nations. Faced with enormous amounts of intelligence, they turned to the burgeoning documentation community to develop distribution systems.[41]

Of course, the United States was not in the war alone, and Anglo-American cooperation was especially significant in intelligence and espionage. The famous "special relationship" between the United States and the United Kingdom was founded in large part on sharing scientific information and wartime technology. In the early years of the war, when the United States was still technically neutral and it appeared that Germany might soon invade the United Kingdom, leaders on both sides of the Atlantic saw the benefits of sharing information. Sir Henry Tizard was key in planning and leading one of the most important of these scientific attaché missions in July 1940. Britain offered advanced knowledge about the emerging nuclear science and about radar technology, sending many of its top scientists (including Klaus Fuchs, the scientist/spy who passed Manhattan Project plans to the Soviet Union) to the United States. The United States offered a safe base of operations, far from German bombs, and vast resources, infrastructure, and political will to pursue military research of all kinds.

In January 1942, the US government became aware of the AMS, which had begun its widespread periodical copying program. In April, to coordinate with this service, the US intelligence community sent Eugene Power to serve as liaison and representative.[42] Power was an expert in micro-reproducing scientific materials, both through his own experience in libraries and as vice president of a private firm, University Microfilms, Inc. In addition to helping the US government produce copies of enemy documents, Power arranged a deal where University Microfilm would receive its own copy of numerous periodicals, which the firm could resell to American industry for profit. In May 1942, the British made the AMS officially interallied by including American representatives on its managing executive committee. By late 1945, this AMS program had processed more than 5.5 million pages of enemy periodicals.

Within the United States, the newly formed Periodicals Reprint Service within the Library of Congress worked to make these materials available to libraries. American copyright sensibilities were not so tender, and the Alien Property Custodian simply seized all Germany copyrights (as was the case during the First World War). Which documents they would reprint was decided by an advisory committee. This committee included eight of the top librarians and documentation experts in the United States, including: E. J. Crane (editor, *Chemical Abstracts*), Watson Davis (president, ADI), Sarah Jones (librarian, Bureau of Standards), Keyes Metcalf (president, American Library Association), Luther Evans (Assistant Librarian of Congress), and Paul North Rice (executive secretary, Association of Research Libraries).[43] The program eventually covered a wide territory, including acoustics, aviation, biochemistry, electronics, engineering, enzymology, explosives, mathematics, pathology, petroleum, plastics, rubber, and virus research. This service soon had more than nine hundred subscribers (many of them from the United Kingdom and its empire), of whom the vast majority were working directly on war-related research.[44]

France's War of Bibliographers

Scientific bibliography was at the heart of a deeply personal, bitter dispute that split the French—and indeed, the international—scientific community in the early postwar years. Unsurprisingly, it centered on the efforts of Jean Gérard.[45]

Within occupied France, access to American, British, French, and Soviet journals dwindled, crippling remaining French science. For Gérard, this was unacceptable, and he turned to the main source of foreign science still available to him: Nazi Germany. Gérard was not alone in this. In October 1940, a new director of the Bibliothèque nationale, L. R. Faij, reported that forty other institutions sought to reopen scientific communication with Germany.

German administrators were happy to cooperate with the French. In German society, science (*Wissenschaft*, though the German term is really much broader than this English translation and includes systematic study of literature, society, etc.) was seen as deeply connected to culture. Spreading German culture, in turn, was actually seen as beneficial to pacifying and controlling a European empire. Promoting German culture was one reason that one of the great German pioneers of documentation, the chemist Wilhelm Ostwald, had pushed for standards in scientific information (and the use of German as an international language of science).[46] It was also one of the reasons that American and British agents had a fairly easy time acquiring German science journals in neutral countries: spreading German culture via science was not seen as a problem, so science was not nearly as closely censored as it was in the Allied nations.[47]

Franco-German cooperation in sharing scientific periodicals allowed French scientists at least some window into international developments. It also gave Gérard's abstracting service, SOPRODOC, a near monopoly within occupied France, further elevating both his leadership role and the importance of the Maison de Chimie. These competing interests were at the heart of postwar debates about Gérard's "collaboration" with the occupying enemy.

The only rival SOPRODOC faced was an illegal service that published abstracts of Western science, operated by the Centre national de la recherche scientifique, or CNRS (the same agency that would eventually run postwar FIAT-like missions in the French zone of occupied Germany). The CNRS had been formed in October 1939, in the early days of the war, to centralize all basic and applied science, in part to allow for more efficient war research. As such, it had little institutional sway before the German invasion and operated primarily in the Vichy state.

Frédéric Joliot-Curie, Nobel laureate physicist and son-in-law to Marie Curie, assigned Jean Wyart, a young crystallographer, to run the bibliographic efforts of the CNRS. Joliot-Curie had extensive connections across both occupied and Vichy France, including an astronomer who received automatic passes across the Vichy boundary line from an officer in the Luftwaffe's Science Service.[48] This astronomer then lent these passes to Wyart, offering him rare freedom of movement between German-controlled and Vichy's nominally neutral territories.

Within Vichy, Wyart had access to a military science service that subscribed to American and British journals, as well as to scientists sympathetic to both the Allied powers and to French science. Wyart's abstracting service, then, provided a rare, illegal glimpse into Anglo-American development and was in high demand. Gérard, in turn, saw this as a threat to SOPRODOC, and perhaps to the Nazi government's relative leniency toward accessing German and Italian science, and sought to suppress Wyart's efforts. When Gérard used his influence to limit CNRS access to already-scarce paper, Wyart turned to a friend at the school of papermaking in Grenoble. Though printing anything required German authorization, a small printer involved in French scientific publishing agreed to produce the CNRS's *Bulletin Analytique*.[49] Together, the *Bulletin* and SOPRODOC covered virtually the entire scientific world, neatly divided along the lines of wartime alliances. Whether by negligence or disinterest, the Germans never ordered Wyart to cease his well-known and popular efforts.

Within a month of Allied forces entering Paris, inquiries began into Gérard's wartime behavior. The CNRS drew legitimacy from its opposition to the Vichy regime, and Joliot-Curie took it over as the new centerpiece of postwar French science. After a meeting of the CNRS leadership, Joliot-Curie informed the Ministries of Foreign Affairs, Interior, and Education nationale about Gérard's collaboration, which was "disastrous for the prestige of French science."[50] Acting on this direction in July 1945, a letter signed by Joliot-Curie (but written by Wyart) arrived at the New York office of Marston Bogert, president of the International Union of Pure and Applied Chemistry, asking for his advice regarding how to deal with Gérard, who had recently been imprisoned for collaboration.[51] Bogert wrote to Gérard in November informing him that he had been removed from the union, but Gérard resisted, refusing to resign and insisting that Bogert did not have that authority.[52]

Gérard, in turn, reached out to his own connections, including Charles Parsons, secretary-general to the American Chemical Society. He insisted that he had done nothing wrong and acted only in the public interest; that he had only ever received warm congratulations for his work during the war; that he had resisted "at great risk" numerous German efforts to transfer the seat of the International Union from Paris to Germany; and that he had only ever acted with "grand impartiality."[53] After fighting for some time and being imprisoned for several months for collaboration, Gérard resigned in February 1946, just in time to avoid Bogert's request to Joliot-Curie that he go through Nazi files to see if Gérard was listed as an agent.[54]

The Dam Bursts: Grappling with Masses of German Documents

The amount of technical information gathered in occupied Germany is staggering. Teams of scientists and industrial technicians determined what information would be microfilmed on-site during visits to factories and plants, then shipped the microfilm to headquarters for processing and publication. In the case of one visit to Degussa (Deutsche Gold- und Silber-Scheideanstalt), this meant about 110,000 pages of material from the main office and two production facilities combined; at the Dr. Alexander Wacker Gesellschaft für elektrochemische Industrie, GmbH, about 20,000 pages.[55] Surveying just sixty-seven plants, FIAT estimated that they would select from thirty-three million pages of material out of three billion pages screened, a process they estimated would take about eleven years.[56] This was deemed improbable, but efforts to limit the scale of document collection to make it possible to digest were only partly successful. The anecdotes of American chemical researchers discussed in chapter 1 tell a similar story, as would the stories of many other teams of technical investigators.

By the end of 1946, the British Board of Trade was relieved to hear that they had halved the estimated amount to be microfilmed—to half a million documents and reports related to German developments. These documents had an average length of ten pages, meaning their new, lower estimate was merely five million frames of microfilm. This, with characteristic British understatement, was "even so . . . more than enough to give us a first class 'headache.'"[57] On the US front, those publicizing the exploitation efforts in such trade publications as *College and Research Libraries* explained delays by pointing to the "tens of thousands of tons of reports and publications" involved. By late 1946, a meeting of the Technical Industrial Intelligence Committee (TIIC) reported the microfilming about half finished, having filmed 775,000 pages of chemical industry documents alone, as well as hundreds of thousands each from other fields.[58]

Taking this 50 percent estimation at face value, the TIIC aimed to film about 7.8 million frames of material. Among these were a mix of scientific, technological, and intelligence assets: "pending patent applications for the war years, doctors' dissertations in the natural sciences and medicine, wartime issues of several hundred scientific and industrial journals which are not presently available in the United States, all technical documents from the German Government ministries," and the documents selected by British and American FIAT investigators.[59] Even discounting the additional material taken in by military intelligence units and other nations' investigators, this would combine to almost thirteen million frames of microfilm. Printed out, this would form a stack of paper almost a mile high.

Once the film reels reached each nation's headquarters, serious problems arose. One major issue was keeping track of what information was on each roll. Few end users were interested in ordering microfilm rolls with hundreds or thousands of pages of files on them, labeled by industry but rarely more precisely than that. Those gathering the documents were theoretically responsible for generating reports about the technologies they investigated, but none attempted item-level descriptions of what they were microfilming. Upon returning home from a trip through Germany, if a former investigator wished to consult a particular document, it might be anywhere on one ream of microfilm—itself containing hundreds of pages of images—mixed in among hundreds of other reels covering the same general subject area.

Another obvious but serious bottleneck was that most of these documents were in German. As of 1958, 49 percent of American scientists claimed to know at least one foreign language (a much greater number than today).[60] However, even if numbers were slightly higher in the mid-1940s, and the foreign language they knew was usually commonly German, most end users would need this material to be translated as well as indexed, selected, copied, and shipped. These document centers then faced an additional set of decisions for every batch: How much time should they spend figuring out what was in each film roll and keeping well-indexed records? Which documents, if any, should they translate? In theory, it would be more efficient for the central bodies to undertake translation, rather than require each end user to duplicate the process, but where would they get these funds?

Even by August 1945, American officials began apologizing that reproduction of reports had "bogged down quite badly recently."[61] Delays on delivering documents ordered by industry or scientific societies grew to longer than a year, leading one trade journal to jokingly celebrate that at least they "will be made available to historians that are active in our grandchildren's time."[62] Initial investigations had found "a great volume of documents" that "soon pyramided to a volume beyond the physical ability of the limited teams to examine."[63] "Recognizing its limitations," FIAT (US) decided eventually to bring all materials to the United States for processing, rather than handling it in conjunction with British efforts, though this "effectively exclude[d] the possibility of British positive-copying the US films and microfilming the cards."[64]

Initial planning between the US Department of Commerce's Publications Board and BIOS indicated that the United States would provide weekly bibliographies for free to American and British industry, from which industry could place orders, with the topics limited to German scientific and technical documents (taken to mean industrial science).[65] Tightening budgets led the United States to change policies to charge for the bibliographies, while their scope expanded to include academic science journals, intelligence reports, and other technical documents—"a complete hotch-potch, not only of intelligence from Germany but also from US, United Kingdom, Japanese, and even neutral sources"—a reflection of the documentation activists' goals of cataloguing all the world's information.[66] As US bibliographies expanded to include all of this new information, they became (in the British's eyes) "completely unsustainable . . . and likely to become even more so."[67] The result was a bibliography from which orders could be placed, but the documents themselves often could not be found. Eventually, BIOS banned the distribution of US Department of Commerce bibliographies within the United Kingdom for this reason. Summing up the British perception of American efforts at organization, bibliography, and indexing, a Board of Trade official explained internally that FIAT (US) did "not see things from the customer's view-point."[68] Technology transfer was not just a problem of having information *available*, but also *receiving* and *integrating* that information, and British authorities increasingly felt that their American counterparts had not thought through both ends.

In response to British complaints, the head of the Department of Commerce's Publications Board, John Green, resorted to more old-fashioned methods. He hired a private "sleuth" to hunt through the growing document collection to find specific items within the IG Farben records, fulfilling orders slowly but (at least in isolated cases) effectively. Green promised that such sleuthing would be possible in future document hunts.[69] Otherwise, he turned to Mrs. Dorothy Gordon, who "had built up this Department from its inception and possessed a quite exceptional personal knowledge of the history of practically every report, and knew into which channels to direct a search for supplementary data."[70] The contribution of such a librarian was invaluable, but the talent of one woman for finding files was hardly an accomplishment of the original goal of a thorough, navigable bibliography of this massive collection of scientific knowledge. The end of duplication with the British, in turn, raised an issue of data vulnerability: if the copy was "ruined by damage in the out-of-date apparatus now held, we shall be completely at a loss to service industry with that particular document."[71]

Alfred King, head of the British Scientific Office, wrote to the head of the German Division of the Board of Trade in September 1946 suggesting a pause in operations while they considered investing in an innovation that "may change our attitude toward the whole transaction" with FIAT (US): "cards designed for mechanical sorting."[72] These mechanically sorted cards had caused quite a splash at a symposium on technical bibliography at the American Chemical Society meeting of 1946, and King immediately saw the possibilities. "The mass of technical material now being accumulated is so great that unless some efficient method for choosing one card out of millions is adopted, the information is not readily accessible and is in the end lost to science and industry."[73] Ultimately, BIOS would go on as it had, but the international scientific bibliography movement—with its focus on new information technologies—continued to shape the expectations and policies of the efforts to exploit German science.

In any event, by the best estimates of the TIIC, even in 1947 there were "literally hundreds of tons of undigested data scatted in a number of repositories in Germany, France, England, and Japan."[74] The United States "admitted that [they] had attempted to publish far too much material," and yet far more remained out there to be collected.[75] Those responsible for leading efforts to exploit German science and technology had adopted the ambitions of the early utopian bibliographers, aiming to gather and index all the scientific and technical knowledge of humankind, and they had adopted its more recent tools in microfilming and distribution. In the face of congressional budget cuts in the late 1940s, these ambitions proved far too lofty to accomplish, and the exploitation efforts—though far from useless, it should be emphasized again, and quite successful in some areas—choked on the glut of information.

When the oil crisis of the 1970s caused some researchers to look back to Germany's midcentury developments, researchers found masses of untranslated, largely untouched files on oil research in American archives.[76]

Intelligence, Secrecy, and Bibliography

Allied exploitation programs faced another barrier that the documentalists did not: secrecy. The documentation movement was based on an ideal of international, open science. In their dream scenario, every document could become immediately accessible to anyone, anywhere in the world. This, of course, would be a nightmare for militaries and intelligence agencies. This conflict frustrated scientists and slowed collaborative research during the war, including on the Manhattan Project.[77] When it came time to process and publish the microfilm reels sent back by FIAT/BIOS investigative teams, this mismatch between documentalist-inspired planning (e.g., bulk-microfilmed files made accessible through indexes, abstracts, and bibliographies, distributed throughout relevant communities, and aided by specialist libraries) and the realities of early Cold War secrecy created another major barrier to the Allied exploitation programs' success. Differences between American and British ideas about classified knowledge also created another significant point of contention between these allies, despite their close cooperation in most intelligence matters during this period.

Both American and British officials complained about the other side over-classifying relatively innocuous material and being too lax about potential threats. Much of the problem originated, again, in practicalities of the processing and distribution of so much information. One American memorandum from the TIIC to the Communications Subcommittee in August 1945 expressed frustration that "a large part of the information contained in the English language reports and other documents marked 'SECRET' or 'CONFIDENTIAL' is in fact not properly classifiable at all. (Some indeed should be marked 'TRIVIAL')."[78] The Aeronautics Subcommittee warned that "any classification on the information you send us slows down the method of distribution," possibly to the point of uselessness.[79] Despite anxious American complaints, the Halstead Exploitation Centre for German documents distributed all of its materials without classification and left security measures up to receiving organizations.[80] This was largely a practical decision, driven by staff shortages and time constraints—few personnel knew English and German, could recognize worthwhile scientific material, knew the guidelines for classification levels, and were willing to work a temporary job of this nature.

Similarly, though the British complained that Americans stamped security classifications on over 20 percent of reports on German firms (including those based in France), they were also quite upset by unilateral American decisions to downgrade classifications. Rather than acquiesce to the US-proposed security classifications, BIOS officials wished "to make it quite clear that as far as CIOS and Evaluation Reports are concerned . . . it is not proposed automatically to agree to the US security classification. On the contrary . . . we propose to retain the existing UK security grading . . . regardless of any re-grading given by the US authorities."[81] Further objections to Americans declassifying reports occurred intermittently from 1945 to 1948. The extent to which this was a serious concern about information leaking to the public is unclear. The British Foreign Office was still quite hopeful of the possibilities of exploitation in 1947. As one official argued: "Prima facie, since the microfilms and prints are advertised as being available without reservation of any kind, we have no real grounds for refusing supply to other nationals. There is, however, a strong feeling that these original German documents and drawings represent a form of reparations—perhaps ultimately the only really valuable form of reparations we shall obtain—and that it would nullify their value if they were to be disseminated to overseas competitors of our own industry. We have tried to meet this situation by administrative delays, thus giving our own industry a flying start."[82] On the American side, the Publications Board of the Office of Technical Services (OTS) within the Department of Commerce fought to reduce or eliminate classification of German documents, though it made little progress. Within the "great reservoir of scientific and technical knowledge . . . developed during the past five years" was surely the "molecules" from which new industries and "jobs for all" could spring, explained the Publications Board to the library and information sciences community, yet this knowledge was being "dammed up by the walls of secrecy."[83] Here, too, qualified staff with security clearance, knowledge of the German language, and scientific or technical training were exceptionally hard to come by, and a great deal of knowledge classified by virtue of coming through intelligence channels never received a declassification review. The declassification of papers from the Manhattan Project led to only a trickle of papers emerging by the end of 1946.[84]

Impact of Documentation on the Exploitation of Germany

The exploitation of German science drew heavily upon the technologies and methods developed by the librarians, scientists, and entrepreneurs who sought to solve the perceived crisis of an overload in scientific information in the early twentieth century. One of the original reasons that the Library of Congress funded microfilm research leading up to the First World War was the need to preserve the priceless rare books and other collections in European libraries and archives. In a sense, the scientific exploitation programs had a similar goal. Both faced a battle against time, hoping to copy and preserve vast quantities of unorganized technical documents before political will faded.

Perhaps the most effective way of showing how deeply the documentation movement, scientific intelligence, scientific societies, and postwar "information science" overlapped is by charting out just a few of the many individuals whose careers touched on each. Allen Kent, one of the most important figures in American information science in the postwar years, moved from the Air Corps to the American Chemical Society to a CIA-funded position at the Massachusetts Institute of Technology, then went on to found an important information school at the University of Pittsburgh, also using intelligence funding.[85] One of Kent's colleagues in Pittsburgh's information science unit, Jesse Shera, had served in the OSS (processor to the CIA), where he consulted on how to use IBM machines in information sorting. Anthony Debons served as a "technology reclamation officer" in postwar Germany before moving on to study experimental psychology with funding from the Office of Naval Research, then moved into the study of scientific computing in the 1950s.[86] This list of examples could be expanded much further.

One of the most prominent scientist-administrators of the era, Vannevar Bush, was also a key link between documentation, exploitation of Germany, and postwar information science.[87] Before and during the war, Bush became an ardent documentation and microfilm enthusiast, using his position at MIT to promote research into machines that could automatically sort through microfilm in search of particular patterns. As the war got under way and Bush took over the Office of Scientific Research and Development, he continued to push a team of his students to work on both a version of the machine to serve the intelligence community's code-breaking needs (the comparator) and the original library-based information retrieval device.

Neither machine was ever particularly functional during the war, and in 1946, Bush turned to John Green of the OTS (whose main responsibilities included publishing FIAT/TIIC reports for US industry) for funding. Bush had been instrumental in establishing the OTS in the first place, and its attempts to create bibliographies of German, British, and other documents gathered by FIAT seemed like an ideal test case for Bush's machines. Bush, Ralph Shaw (of the US Department of Agriculture Library), and other documentation specialists played a key role in shaping information policies in the OTS, and, in the words of historian Colin Burke, "Shaw's experience with microfilm and his background in technical indexing proved central to Green's drive to turn the war into a scientific windfall for American industry."[88] Bush's microfilm-based machines never functioned very well, and Bush himself was more inspirational than actually foundational in the documentation field that would become known as "information science" in the early postwar years.

There were many similar, interweaving connections among libraries, information science institutions, and FIAT-related agencies. The US Library of Congress turned itself into a war-related research hub during the war, drawing upon special librarians to organize a top-secret research branch for the OSS and applied science units for the military. After the war, it and the Department of Agriculture were instrumental in distributing FIAT/TIIC reports, including coordinating with the Department of Commerce on distributing to appropriate industry groups.

Ironically, this expertise and (at the time) high technology being brought to the task might have undermined the entire affair. Microfilm was a wonderful technology for *capturing* and *transporting* vast quantities of information, and a workable one for *copying* it. This ease, combined with time pressures and the impossibility of having expert screeners in every field, led to teams gathering enormous quantities of information of varying quality and trusting in the indexing/bibliography/abstracting process to make them useable and useful down the road. For all the optimism and grand planning around punch cards, card catalogs, Dewey decimal and other bibliographic organizational schemes, and efforts to establish standards for coordinating international approaches to these topics, they were not up to the task of taming wartime German science.

It is worth pausing to consider alternative forms this project might have undertaken. One option would have been to seize the original documents from Germany rather than create duplicates. This would have been tremendously disruptive to German industry and thereby to the recovery of the European economy as a whole. Without test data, blueprints, or general records, the only remaining embodiment of these technologies would have been the remaining workers (many of whom had died in the war, been seized for forced labor by invading Allied forces, or left to scavenge during the postwar food shortages). The destruction of scientific and industrial libraries—looted to some extent anyway, especially by French and Soviet authorities trying to build up their own recently looted collections—would have been a major impediment to rebuilding German science. These papers would have filled about three thousand banker's boxes, weighing about seventy-five tons, and so would have been a major logistical challenge—and could not have been shared among the Allied powers, further exacerbating diplomatic tensions.

Microfilm, then, did accomplish a minor miracle. Teams of investigators could set up a mobile microfilming team on-site for larger collections, or temporarily borrow smaller collections to process them at document centers before returning them to their owners. While not free, duplicating these film strips was reasonably economical. Storage needs were about 5 percent of the paper equivalents and weighed proportionally even less, which was useful for air transport.

It is possible to imagine another alternative where investigators gathered relatively few documents. The *FIAT Reviews of German Science* accomplished a great deal in terms of catching academic researchers up on wartime advances, and normal avenues of scientific publishing (e.g., journals, occasional books, conference talks, etc.) could have fleshed out these introductions. In terms of industrial science and technologies, sending industrial engineers simply for their own benefit might have had the most impact. British attempts to group together investigators along industry lines made sense in principle, though they ran into issues of the groups then having internal rivalries. Ultimately, whether this would have been an ideal path is impossible to know. Regardless, the ambitions and methods of 1940s "information technology," and faith that these technologies could solve what turned out to be (at least in part) social problems, deeply shaped the investigations of German science.

Bibliography and Scientific Communication in the Cold War

While documentation (and connected movements) did more to shape FIAT than FIAT did to shape information science, the investigations in Germany did have some meaningful legacies for international scientific publishing. As the Cold War escalated in the late 1940s and set in over the 1950s, science took on enormous importance for national prestige in addition to its practical contributions to national defense. In an increasingly polarized world, science information systems and publishing venues became pawns in the diplomatic struggle.

One profound—albeit indirect and unintentional—legacy of FIAT investigations was starting the transformation of international science from polyglot (i.e., with publications balanced among English, French, German, Russian, and then some regional publications in other languages) into the nearly monoglot, English-dominated world of science we have today. This came about, as historian Michael Gordin has argued, through the development of science translation services that made nearly everything available in English—and these translation services, in turn, received vital early funding and attention due to the need to translate masses and masses of captured German science.

The Publications Board of the US Department of Commerce began recruiting companies, trade groups, and scientific societies in July 1948 to scan through microfilm sent from Germany and write out abstracts, bibliographic entries, and other metadata for use in processing.[89] Budget cuts meant that the only compensation offered was "being the first to scrutinize the material" (other than those who had screened and microfilmed it in Germany, presumably). At least some firms accepted this offer, sometimes also borrowing Department of Commerce microfilm-reading equipment. As expected with a volunteer effort, though, results were spotty. Industrial users with real interest had to order large collections (since they were rarely well-indexed) and pay for translation.

One entrepreneur, Earl Maxwell Coleman, founded a profitable business called the Consultants Bureau on the realization that (as he estimated) "twenty-one tons of captured German war documents" were being retranslated by each individual customer who ordered a copy.[90] Finding that the American Petroleum Institute alone had one hundred microfilm reels of one thousand pages per reel in need of translating, Coleman's business was one of the first in an industry of mass translation. He hired dozens of German-speaking, technically capable translators, most frequently immigrant academics in need of extra income. In Coleman's recollection, "German died on the vine . . . American companies had already acquired the know-how laid out in the documents."[91] The basic business model was sound, however, and the Consultant's Bureau shifted its focus to translating other types of scientific and technical documents, including journals and book-length manuscripts.

The Consultants Bureau and similar firms dramatically lowered the cost of translation. This, in turn, made it much easier for English speakers to get by without learning foreign languages or, if they were native in another language, the need to learn just one (English) rather than many.[92] As American science increased in prestige and volume, these English-language scientists read foreign journals in their own languages less and less often. An article published in German or Spanish would therefore receive substantially less attention than one in English, regardless of its merits. Once a critical mass of international scientists learned English, it became self-reinforcing: native English speakers had less and less reason to learn other languages, and those outside the Anglophone world had all the more incentive to learn and use English.

Within the world of library science and information science, the hodge-podge of documents included in FIAT microfilm rolls (journals, research notes, patent documents, etc.) contributed to a key challenge for special libraries in the postwar years: sorting and storing technical reports. These "white papers," "grey papers," and other research reports lacked the formatting standards and clear authorship of books and traditional periodicals, so they did not fit traditional bibliographic schemes. How could you file a report written by a team of dozens or hundreds of researchers, unpublished but still valuable? My claim is not that FIAT was the primary driver here—if any one institution deserves credit for innovating special library techniques for these materials, it might be the US Atomic Energy Commission's Technical Information Service, built out of the Manhattan Project. Still, these FIAT-related collections were one of the items in mind when the Central Air Documents Office, Library of Congress, National Advisory Committee for Aeronautics, and Atomic Energy Commission came together in 1950 to establish rules and procedures for processing these types of materials in library settings.[93] Similarly, the British Library and provincial libraries throughout the United Kingdom were the primary mechanism for distributing BIOS reports to smaller firms and the public, generating complaints and questions about how to fit them into existing filing systems.

Robert V. Williams, writing on the history of the special libraries and documentation movements within the United States, argues that "in the limited pre-war activities [of US groups] there was a definite orientation towards large-scale dissemination projects using microfilm. Postwar activities, however, showed signs of more detailed focus on the organization, control, and use of scientific documents."[94] The period of "information turmoil" generated by "a tremendous increase in the volume of scientific information, particularly in the form of technical reports," was a major cause of this shift. Irene Farkas-Conn, in her history of the American Society for Information Science, reiterates this argument.[95] During the war, the massive amounts of information gathered by the air force, OSS, and other forces (presumably including FIAT) led these agencies to enlist Aslib and the Library of Congress in micro-filming efforts, resulting in 300,000 Japanese and "13 million pages of censorship intercepts on microfilm [being] deposited in the National Archives under the presidential seal."[96] Dealing with this influx of material, bibliography efforts to make material available to the intelligence community, and a mandate from the assistant secretary of state to reduce duplication among these special libraries meant "for the first time US librarians examined from a broad point of view how the research libraries could best serve US scholarship and developed a national plan."[97] These challenges also led to an expansion of early computing technology.

Scientific Bibliography and Cold War Rivalry

Denying rival powers access to German scientists and science was a key motivation for both Western and Soviet planners, and this logic extended into Cold War science information systems. As wartime alliance gave way to the Iron Curtain in the late 1940s, American and Soviet governments came to see organizing science within their "bloc" as a symbol of power and tool for influence. This meant both co-opting German scientific publications and extending their own networks as far as possible.

The Soviet Union was particularly concerned with controlling the flow of scientific information for ideological as well as practical reasons. Scientific and technical elites, and the institutions they ran, had been necessary allies during the Soviet Union's early economic struggles and then the war, despite their suspiciously internationalist, bourgeois backgrounds. As Nikolai Krementsov notes, the wartime alliance broke down barriers between Soviet and Western scientists in the late 1930s. Stalin even invited scientific delegations from Allied nations to a June 1945 celebration of the Academy of Sciences of the USSR.[98] By 1947, however, as tensions rose over plans for Germany, the Soviet government cracked down on Soviet scientists using (or even referring to) Western science, traveling to conferences without secret police escorts, or publishing their research where Westerners might have access. Instead, Stalin and subsequent Soviet leaders sought to build a self-sufficient, superior world of science within the Soviet sphere.

One path toward Soviet bloc leadership was to draw from German successes. From its founding in 1830 through the Second World War, the publication *Chemisches Zentrallblatt* was a tremendously important abstracting journal in chemistry, used around the world.[99] In a sense, its overwhelming size was an embodiment of the information problem of the twentieth century, as it sprawled into thousands of pages. The American Chemical Society's *Chemical Abstracts* represented a kind of competitor starting in 1907, but *Che-*

misches Zentrallblatt remained vital right through its press being bombed into rubble in 1944.

As the Allied powers debated how to deal with occupied Germany, and eventually moved toward divided East German and West German states, both sides tried to revive the *Chemisches Zentrallblatt* as a symbol of their legitimate ties to Germany's more honorable past. The German Chemical Society, which published the journal, was itself under unknown control in divided Berlin. American forces had moved the editorial offices of the German Chemical Society west to Dahlem and made an effort to restart the journal, but the Soviet Union was insistent that it be an East German accomplishment and gathered former contributors to put together a competing edition. Much as American forces were willing to take German scientists who sometimes had clear Nazi pasts, Soviet authorities decided that the former editor of the *Chemisches Zentrallblatt* should be forgiven for having joined the Nazi party in 1933, if it meant greater legitimacy for their edition. Eventually, a compromise between the sides allowed a 1950 edition to be jointly published by the East-West German Chemical Society.

Within the Soviet Union, the Academy of Sciences created a new scientific information clearinghouse in 1952, dubbed the All-Russian Institute for Scientific and Technical Information (VINITI).[100] The war—and the massive amount of information brought east from Germany—had been a major instigator of reform in information management within the Soviet military, much as it had been in the Western Allied nations. Among other coordinating activities, VINITI was to publish an abstracts journal, covering international science and technical literature. This was a massive enterprise, of course, and one ultimately not as well-funded as later American postwar abstracting services. Still, it was fairly revolutionary for its time and covered 1.3 million scientific publications worldwide by 1990.

In the United Kingdom, John Desmond Bernal attempted in 1945 to orchestrate a nationalization campaign to centralize scientific abstracting, bibliography, and publishing. His earlier connections with the Communist Party and trips to the Soviet Union proved to be greater obstacles during the early Cold War era than they did in the 1930s when he promoted Aslib and other bibliographic efforts, however. When Bernal proposed the formation of a National Distribution Authority for Scientific Information, the scientific societies and individual scientists objected to the proposal's threat to scientific freedom but also to the politics of the proposer. Bernal was dubbed "the well known left wing planning enthusiast," and his plans—though quite in line with the 1920s documentalist dreams of rationalizing and centralizing information— were now seen as politically dangerous.[101]

Instead, British government involvement in scientific communication had a natural home in the DSIR, which preexisted the Second World War, and in its wartime centralization of information into the Science Museum as a de facto national science library.[102] The DSIR's Information Division absorbed the bureaucratic remnants of the Board of Trade's processing of German information, and in 1955, it began planning a national lending library for science, opened in 1962 at Boston Spa in Yorkshire.[103]

In the United States, the government took a far greater interest in science abstracting and publishing for at least one important reason: it was now funding most science, through the military or Atomic Energy Commission (itself heavily spending on nuclear science to advance weapons technology). The story of American funding for rebuilding European science, and the diplomatic power that it gave American leaders in a tense Cold War atmosphere, has been best told elsewhere, in particular in Krige's *American Hegemony and the Postwar Reconstruction of Science in Europe.* Those programs very much involved science publishing and communication systems.

Within the United States, the bureaucracies that ran exploitation of German science evolved directly into government-run science information programs throughout the Cold War.[104] The OTS had been assigned to run the Publications Board of the US Department of Commerce. In 1965, it expanded into the Clearinghouse of Federal Scientific and Technical Information, which in 1971 became the National Technical Information Service. Each of these units was responsible for providing summaries of foreign technical reports to industrial subscribers and research institutions. The air force document processing facilities in London and at Wright Field eventually merged into a Central Air Documents Office, which became a unit of the Armed Services Technical Information Agency, then was renamed the Defense Documentation Center in 1965. In 1958, the President's Science Advisory Committee investigated whether such an agency should be created in the model of the Soviet VINITI but recommended instead strengthening the scientific information distribution role of the National Science Foundation.[105] The National Science Foundation's Office of Scientific Information Services resulted from this recommendation.

The problem of information exchange in the ever-growing sciences became a central issue in the newly important field of science policy. Vannevar Bush, former head of the wartime Office of Scientific Research and Development—and whose recommendation had spurred President Truman to issue Executive Order 9569 creating the Publications Board—highlighted this problem in his famous policy statement, *Science, the Endless Frontier*.[106] "International exchange of scientific information is of growing importance," he wrote, and he saw a role for an active government in pursuing it via "the arrangement of international science congresses, in the official accrediting of American scientists to such gatherings, in the official reception of foreign scientists of standing in this country, in making possible a rapid flow of technical information, including translation service, and possibly in the provision of international fellowships."[107]

Conclusion

There are countless ways in which the politics and diplomatic tensions of the era (including in nonaligned countries avoiding allegiance to either superpower) reshaped science information systems and scientific publishing, but much of that history lies beyond the scope of this chapter. In the following chapter, I broaden the question of the long-term legacies of the exploitation of German science, taking a bigger-picture approach to ask how it fits into these stories of Cold War science and diplomacy, technology transfer, and business history. Here, the most important point is that the exploitation of German science—though unprecedented in many ways—cannot be understood without this longer history of ambitious efforts to make all of the world's science free and freely available to all, leveraging information technology to create searchable, accessible databases.

The linkages between the development of information science (growing from the documentation movement) and the scientific exploitation efforts in Germany are rarely necessary-and-sufficient causal relationships, yet they are pervasive. You could explain the FIAT Reviews without these links; you could find a reason why files about Jean Gérard's postwar tribulations are scattered throughout the French FIAT records; you could write off the spur to commercial machine translation of scientific texts as simply an idea and technology whose time had come. Certainly, the FIAT-like efforts were hardly the *only* cause for intelligence community investment in mechanical sorting (and eventual computer encoding) of bibliographic entries for information retrieval or increased state attention being paid to funding abstracting services. Yet all of these make far more sense when pieced together.

FIAT and others were neither the origin nor the endpoint of these broader trends, but they were a junction through which the trends passed. Due to the unprecedented extremity of the information problems FIAT faced (trying to handle not just the incredible amount of scientific information already problematic before it piled up for several years but then also all of the assorted information of potential intelligence value beyond that), it served as a catalyst for shaping the future of international structures for scientific communication. When military intelligence agencies targeted scientific and technical information during the occupation of Germany, they were dealing with preexisting problems and drew upon preexisting solutions. Among these were using the international network of scientific-technical specialist libraries to distribute abstracts and bibliographic information, and depending upon those working in these institutions to contribute their time toward this end. Also borrowed was a faith that through microfilm, all collections are possible.

The growth of information technology in the decades since the Second World War raises an interesting (if fundamentally unanswerable) question: If the occupation of Germany happened with today's information technology available, would the results have been significantly different? Certainly, modern scanning techniques and optical character recognition would have made the results more searchable, and digital technology would have allowed storage and searching to be less of an issue. Publication would not even be a problem, per se, if the resulting information could be hosted on the internet. This assumption might still be naive, however. Much of the most important knowledge acquired during these investigations came from hands-on visits to factories and inspiration for new products rather than copying German developments, and neither could easily be digitized. Further, the greater tools for search would have to deal with massively more information—a problem all too familiar to those struggling with yards of boxes of archival files—much less the quantity that passed through FIAT's now-obsolete system. Nor should we discount the human element—the inefficiencies of processing massive data through military intelligence channels and filtering important information from noise. Information science has undoubtedly advanced since the 1950s, but technology alone may be no more a solution for technology transfer today than it was sixty years ago.

7

Legacies of Intellectual Reparations Programs

Industrial Know-How in the Postwar World

While the core of the intellectual reparations programs took place within just a few years in the late 1940s, their legacies played out for decades. Some of these legacies appear in previous chapters, such as changes in scientific communication. Possibly the most important set of legacies, though, centers on the issue of moving industrial technology across national borders. From the end of the Second World War throughout the Cold War, both private business and governments invested far more heavily in international technology transfer—working to make it happen, profiting from it, and sometimes trying to stop it. This, in turn, meant that the lessons they took away from their efforts in Germany played an important role in shaping the Cold War.

The exploitation of German science was not the only source for these long-term, big-picture trends, but its scale, ambition, and challenges fed into them, reshaping how a wide range of people at the time thought about technology transfer. Conversely, because these trends are broader than FIAT, BIOS, and so forth, they can help us better understand how those programs fit into the story of the Cold War world.

One of the biggest changes, seen again and again in this book, is an international phenomenon beginning right around the start of FIAT, which went on to sweep the worlds of business and policy: a sudden fascination with the importance of "know-how." With hands-on skill that resisted codification, the accumulated minor insights and innovations that were sometimes too minor to be patented yet still mattered, what would later be called "tacit knowledge," businesses now clearly saw that this human element of "the 'know-how'" was fundamental to their technology planning. In the exploitation of

Germany, investigators from around the world and across nearly every industry came face-to-face with the impossibility of putting everything needed into writing. In the postwar world, this focus on the know-how element took on a life of its own, shaping business, law, and politics around the world.

Contracts to share and exchange know-how between firms became commonplace by the 1950s. Groups such as the International Chamber of Commerce lobbied for new legal protections for know-how as an intellectual property right (akin to trademarks or copyright) around the world. Interest in know-how became so widespread that it shaped entire linguistic trends across the globe, as *das Know-how* and *el know-how technico* became a shared terminology in both developed and developing nations. Lawyers struggled to define the term in language precise enough for licensing deals, and government took note. In this chapter, I lay out some of the ways this know-how phenomenon built upon wartime experiences in Germany in the international business world, as well as its consequence for international business.

Another legacy of these programs, built in part on this know-how phenomenon, was a new commitment by governments around the world to building alliances and providing economic aid through sharing technology. In the exploitation of Germany, this meant working together to extract German industrial techniques, then sharing the proceeds with others via written, easily reproducible reports. As this report-writing failed, though, policymakers learned to focus on the know-how dimension through encouraging the circulation of experts rather than (just) documents and blueprints.

For the United States, this meant building up Western Europe by sharing American production know-how while sending "technical missionaries" to developing nations around the world. Within Western Europe, both Britain and France worked to build new kinds of connections with German industry at the same time they absorbed what they could of American methods. The Soviet Union, meanwhile, worked to forge its own sphere of influence in Eastern Europe and eventually China, and sharing industrial technology became an important part of that work as well. Transnational flows of industrial technology became a key part of international diplomacy.

Just as Operation Paperclip and other programs aimed to both learn from German technology and deny it to others, refuting technology to rivals took on broader importance during the Cold War era. The United States in particular expanded its ambitions of denying key advances to the Soviet bloc. One result was the Coordinating Committee for Multilateral Export Controls (CoCOM), a "gentleman's agreement" among the United States and Western

European nations to restrict what kinds of technology could be sold or licensed across the Iron Curtain.

While this book generally focuses more on industrial technologies than military ones, which generally receive more attention, nuclear power production is worth exploring as a special case that shows how each of these trends combined to shape the Cold War world. Denying nuclear weapons technology at first centered on controlling "nuclear know-how," though political forces eventually reduced the influence of this know-how phenomenon in the nuclear realm. Despite the concern of nuclear weapons proliferation, sharing nuclear technology also became a key part of Cold War diplomacy.

This chapter, then, draws together a number of stories that highlight different ways in which the experience of taking Nazi technology reshaped the political economy of the postwar world. The connections are not always direct, nor is the claim that these broader trends only existed because of efforts to take German technology. The longer-term view of technology transfer in Cold War business, diplomacy, and culture—influenced in a meaningful way, at least, by the exploitation of German science—nevertheless gives a different kind of answer to the key question of almost all historical research: Why does this matter?

Private Industry's Big Lesson: The Necessity of Know-How

Throughout the Cold War, a number of surveys asked companies what they saw as the most important factor in technology transfer, and the consensus was clear: "know-how."[1] In 1955, the International Chamber of Commerce argued that know-how had "become in recent times [a] tremendously valuable [subject] of industrial property."[2] The author of a 1953 article in the trade journal *Chemical and Engineering News* argued that a public poll would show that Americans saw the nation's security as based in (1) the atomic bomb and (2) the "great American Production Know-How."[3] From the late 1940s through the 1970s (and, in many ways, through the present day), businesses around the world suddenly came to a collective realization that the human component of technology, what they called "know-how," was invaluable. The result was an international, far-reaching craze for the technological/industrial know-how that spanned the Cold War era.

The intellectual reparations programs in Germany were, at minimum, one key starting point for this phenomenon. These programs caused thousands of businessmen around the world to grapple with the difference between having hands-on, skilled experience with a technology and simply having written descriptions (a distinction later described in academic philosophy as "tacit knowledge" and "explicit knowledge").[4] Over and over again, investigators from these very different countries tried and failed to take Nazi technology, and time and again they described their problem in the same way: the difficulty of capturing "know-how."

The basic idea that there is more to technology than patents and technical documents was nothing new in the 1940s, but with this phrase available to capture and communicate the idea more easily, it took off right alongside FIAT. German firms, as the world soon discovered, retained much of their value in postwar international markets, despite all of their "secrets" being exposed. They retained know-how that could not be microfilmed or packed into crates. At the same time, American, British, French, and other firms came to realize that their own accumulated know-how was tremendously valuable, and a vibrant market emerged of companies licensing their know-how to others. This trend, in turn, generated a whole set of practical concerns for businesses, policymakers, and judicial systems around the world. For example, what kind of legal protection could there be against theft of something as nebulous as know-how?

One demonstration that something was truly new here comes from Google's Ngram Viewer, which charts how popular selected words were as a percentage of all words used in books Google has scanned, covering centuries. You cannot rely on n-gram graphs by themselves, and changes in language do not necessarily mean much anyway. Yet if you look a bit closer at how people were actually using this term in this era, it becomes clear that this was at least seen as something new and vitally important by people at the time.

The FIAT, BIOS, and other records contain many detailed discussions of what know-how meant and why it mattered, and often cast it as a new or emerging category. Frequently, reports include language such as that from a secret report from the Joint Intelligence Committee of the US Army in October 1944: "No effort has yet been made to obtain the so-called 'know-how' which can only be obtained on the ground by qualified engineers and technologists working directly with the military forces."[5] The scare quotes around the term are common, as is referring to it as "so-called" know-how. A TIIC report in 1945 states its mission as gathering "industrial 'know-how.'"[6]

Lucius Clay, the governor of the US zone of occupation, argued to the Quadripartite Coordinating Committee in December 1945 that "scientific knowledge, patents, technical processes, and general 'know-how' constitute a large part of Germany's economic potential and are properly included in the cate-

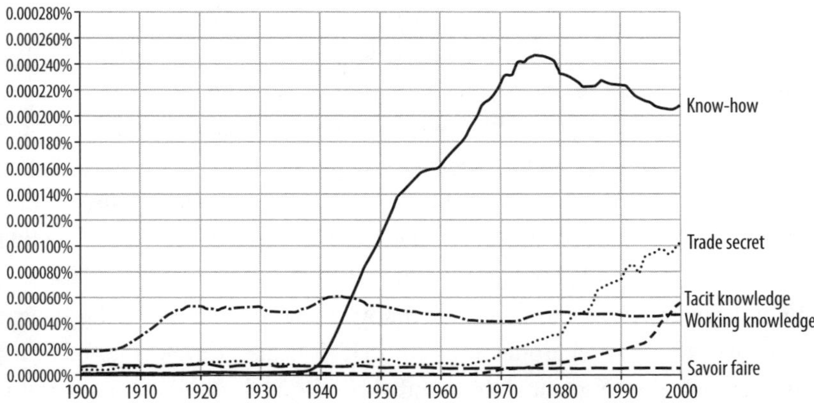

Figure 7.1. Usage of terms "know-how," "trade secret," "tacit knowledge," "working knowledge," and "savoir faire" within the English language as a whole, 1900–2000. Data and graph generated by Google's NGram Viewer, available at https://books.google.com /ngrams.

gory of reparations. . . . This category [of know-how] . . . is important to each of the Allied Powers, but its intrinsic value is indeterminate since it depends upon the extent of parallel knowledge existing within the confines of the respective home government."[7] One set of British planners initially saw reportwriting as a useful scheme because, as they heard from industry: "In this way the German manufacturing process and the 'know-how' surrounding it become an invisible German export, payment being made to the Treasury which credits the sum to Germany's account offsetting the costs of occupation."[8]

To be sure, the sudden business interest in know-how starting in the 1940s was broader than FIAT. An article in the *Wall Street Journal* in 1944 on trends in industry and finance noted:

In drawing contracts for post-war goods the Russians often include clauses permitting their technicians to observe output: they are willing to pay extra for this privilege. Other "technical assistance" contracts provide that American technicians accompany machinery to Russia, to teach workers there how to use it. The British, Dutch, Chinese, others display similar interest. "They are after our know-how," one industrialist explains.

This eagerness to learn US methods has irritated some American businessmen, who fear the loss of trade secrets to foreigners under the guise of war necessities. Most companies cooperate, however.[9]

An article in the trade journal *Electrical Engineering* in May 1949 drew attention to how "American engineering know-how" was an "invisible export," functioning though "license and technical assistance agreements between American and foreign firms," closely aligned with the objectives of the Marshall Plan.[10] In the trade journal *Chemical and Engineering News* alone, articles in the 1940s and 1950s included headlines such as "Know-How, Our Maginot Line," "Foreign Markets for US Know-How Greater Than Ever," "United States Shares Chemical Know-How with the Other Americas," and "Capitalizing on Foreign Know-How," among many others.[11] In the last of these, written in 1956, the author argues that "the US can also develop its technical know-how by adopting what has already been done in foreign countries—in short, by swapping experience with those countries." As his primary example, the author cites his firm's chemical technology as having been "essentially completely developed in Germany with the assistance of German technical personnel."[12] Hiring German technicians in the late 1940s and early 1950s was apparently vital in transferring this technology.

If this phenomenon began in America, it spread quickly. As early as the late 1940s, British members of Parliament debating patent reform commented on a new trend toward the importance of know-how. Viscount Swinton noted in April 1948 that "it is interesting to observe that in the United States the tendency to seek protection by patents is, I think, rather falling into desuetude. The tendency over there, when a new process is discovered, is now not to take out a patent to protect it but rather to rely on being first in the field and having the know-how." Lord Chorley agreed: "In many industries the isolated invention is of no importance. Invention is a matter of building up an immense expertise and 'know-how' and that is a matter the Corporation will have closely in mind."[13] Prime Minister Clement Attlee saw "what is sometimes called the industrial 'know-how'" as the key to preserving atomic secrecy, and Sir James Hutchison later promised to make available "that American jargon—the 'know-how'"—to other European countries "on reasonably favourable terms."[14]

Business and political leaders in the 1950s certainly saw know-how as a new phenomenon, an invaluable business asset, and therefore a potential national security risk. The International Chamber of Commerce launched a study on the "Protection of Know-how" in 1955, concluding:

> Technological improvements . . . commonly referred to as know-how, have become in recent times tremendously valuable subjects of industrial property

supplementing patents and other rights, have assumed a great economic importance, and are the subject matter of an increasing number of very important agreements between business enterprises. . . . Hardly any country so far has dealt in an adequate and comprehensive way with the protection of industrial know-how, although existing national laws on contract, breach of trust and unfair competition are sometimes applicable to the subject.[15]

Nor was the trend short-lived. The journal of the United States Trademark Association noted in 1964 that "know-how is a subject of increasing importance in international agreements and international investment. . . . It has come to be the handmaid of progress and the core of industrial competition."[16] Another legal scholar in 1967 agreed: "There is a current and real interest in the licensing of know-how or technical information. The volume of such licensing is said to be increasing, particularly in dealings abroad."[17]

Know-how licensing became a major American export. The US National Industrial Conference Board estimated total foreign licensing royalties at 500 million USD in 1957, "of which know-how licensing undoubtedly constituted a substantial portion."[18] In a 1958 survey, US firms reported that revenues from the licensing of industrial processes was becoming a major export, with much of the increase driven by foreign demand rather than by their own efforts: "foreign companies seek know-how licenses because of the complexity and expense of obtaining know-how on their own."[19]

The United States certainly had an advantage in selling its know-how in that it had an international reputation for technological excellence, aided in no small part by the war, but other nations also eagerly added to their exports by selling what they could. On its front page in October 1947, the *Wall Street Journal* informed readers of an opportunity: "Know-how for Sale: British Offer US Firms Their Industrial Secrets in Exchange for Dollars."[20] Similar articles discussed Belgians, French, Russians, and many others seeming to buy and sell know-how licenses to US firms during this time period.

German Industrial Know-How in the Postwar World

A good example of how know-how mattered to postwar private industry is the resurgence of the German chemical industry. In retrospect, Germany's great lead in chemical technology was partly an illusion because it was built on a technology—coal-based chemistry—that turned out to be less productive than petroleum-based alternatives in the postwar years. This, combined with the especially heavy attention the chemical industry received in the reparations process (both physical plant and intellectual reparations) and the focus on dismantling chemical plants as having war potential, might well have crippled German chemicals in international markets. Yet while they were no longer as influential as American firms in some ways, the German chemical industry rebounded astonishingly quickly. This was, in large part, because firms around the world still valued the know-how that resided only in German chemical workers.

German firms were the masters of transforming coal tar into pharmaceuticals, rubber, fuel, dyes, and a wide range of other products. With abundant coal and limited access to petroleum, Nazi planners doubled down on the chemical industry's potential to produce substitutes for scarce rubber, oil, and other crucial wartime goods. As a result, this strand of chemical research thrived under the Nazi regime. The prewar American chemical industry was not as powerful as the German competition, especially on the international market. Many firms were in thrall to IG Farben through licensing and patent-sharing agreements, or even directly owned by IG Farben. Partly because German firms had such success in developing and patenting coal-based chemical techniques, American firms invested in using petroleum as a "feedstock" instead. Many of the same products could be made starting from either coal or petroleum, with the difference being in how many steps were required to get from A to B, and how expensive those steps were. There was no obvious inherent superiority to one over the other.

While the Nazi regime insisted on heavier investment in coal-based chemistry, American firms invested heavily in wartime research into petroleum products. They built enormous new plants, developing new processes for mass-producing explosives, fuel, rubber, and other resources from (primarily, but certainly not exclusively) petroleum-based chemistry. This proved to be a major advantage for American firms, in part due to these investments and in part due to changes in the types of products most profitable in the international markets of the Cold War. After the war, the occupying power split IG Farben into several firms—Bayer, Badische Anilin- und Soda Fabrik (BASF), Agfa, and Sanofi being the primary ones. While each successor firm faced unique challenges, they shared two main goals: to overcome their technological backwardness in petrochemistry and to reestablish their overseas market share.[21]

One of the long-term legacies of FIAT was building (or rebuilding) business relationships between German and Allied companies. With this greater knowledge of what firms in each country could offer one another, German chemical firms took initiative in partnering with American firms to exchange know-how. As early as 1948, a group of top managers from Bayer arranged a tour of DuPont and other American firms. Trips like these multiplied rapidly in the 1950s, as more firms had funds available for similar trips. As deals to exchange know-how were struck, teams of German personnel toured American factories and American engineers were implanted in German factories.[22]

Though the flow of technology was now generally from the United States to Germany, this was not entirely one-sided. As historian Raymond Stokes has shown, "many American firms were at least as interested as the Germans in gaining access to technology and know-how from any cooperative agreements with their counterparts."[23] This was something German industrialists realized at the time and used to their advantage. As one Bayer employee reported after a trip to the United States in 1952: "We have to catch up in several areas, but the impression we got is encouraging, too, since we can look with satisfaction on the fact that we are, in several new areas, on the best possible way toward again establishing a starting point for technical development."[24] Coal-based chemistry remained more efficient for producing certain products well into the Cold War, and these mutual know-how licenses allowed West German firms to make a gradual, effective transition, rather than a headlong, frantic rush, to petrochemistry.

A similar overall story can be told about the German rubber and tire industries.[25] Before the Second World War, German synthetic rubber technology was already world class, and the Nazi push for a self-contained economy led to important advances during the war. However, American synthetic rubber was not so far behind. In 1941, prompted by wartime necessity, four rubber firms (Goodyear, Goodrich, Firestone, and US Rubber) and two oil firms signed a pact with the Office of Rubber Reserve to pool all patents and know-how related to artificial rubber. In exchange, the government would fund enormous factories and expanded research.[26] This cooperative agreement worked well enough during the war, but each company was concerned with the long-term viability of artificial rubber once the war ended and East Asian rubber plantations reopened. They hoped to find major advances through investigations of German synthetic rubber technologies and thereby retain economic viability.

Investigations conducted by FIAT and BIOS produced mixed results: "Various innovations . . . generated great enthusiasm among American and British observers alike. It might very well be the case that no fundamentally new methods have been developed, and Germany does not seem to have come so far in comparison with Anglo-American advances, but there is an abundance of individual innovative items nevertheless."[27] Still, these investigations built personal relationships (if strained ones) between investigators and those interrogated, giving each insight into the other's strengths and needs. American and British investigators were deeply impressed with German rubber researchers' know-how, even if not always their physical production infrastructure. German firms, in turn, saw much to gain from longer-term technology licensing agreements with their then-occupiers. In the words of an executive from one German firm, Continental Tire: "We do not wish under any circumstances to pursue courses of our own, but rather orient ourselves to the trails blazed in the tire sector in the United States."[28]

As a result, two major German tire firms, Continental and Phoenix, signed contracts with US firms to send teams of their employees and managers for extended stays in American plants. Over time, it became clear that American technology was not perfectly suitable to German markets. Road conditions were different, Americans were relatively less price conscious but more concerned with a smooth ride, speed limits varied, and adapting existing German factories to new methods required improvisation.[29] These patent and know-how-sharing agreements required substantial investments in research and development to make the technology transfer work in practice. Still, both sides expressed satisfaction at what they gained from the deal. While American technology generally led the way by the 1950s, there was still strong demand for German expertise.

Another example of a company that arguably benefited from occupation-era investigations is BASF, one of the chemical companies formed out of IG Farben. In 1949 and 1950, BASF representative Bernhard Timm visited the United States to tour Dow Chemical and other US chemical industry facilities. There, and on future visits in the late 1950s, Timm was able to form an enduring partnership with Dow in which BASF's most impressive offering was its accumulated know-how—not patents, manufactured goods themselves, or trademarks.[30] Soon, BASF was even able to negotiate Dow out of partnering with Hüll Chemical Works on a joint venture in acrylics by offering to transfer additional know-how at a reduced licensing rate. Besides Dow, BASF representatives visiting America in the 1950s found a welcome reception from many American chemical firms. Many of these business deals were enabled by personal connections between the BASF engineers and the American technical investigators who had spent time in occupied Germany.[31]

This partnership between BASF and Dow was not envisioned as a one-way track. Like firms across Europe and around the world, BASF saw American technology—and the government-industry–academic research infrastructure that made it possible—as the gold standard in many ways and hoped to acquire useful aspects of it through joint ventures. When Dow proposed a new venture to spin synthetic fibers in an American plant, BASF management worried "whether we could supply the required key personnel in numbers great enough to really gain transferable spinning experience."[32] "The know-how expected [to be gained] from forward integration in the USA" was a key metric for many like-minded firms and explains some of the popularity of know-how licenses in this era. Independent of policymakers in Washington or Bonn, private industry worked from the 1950s onward to enable and control flows of industrial technology, and know-how lay at the heart of their thinking.

America's chemical industry was not alone in signing know-how agreements with West German chemical firms in the 1940s and 1950s. The first petroleum feedstock plant in Germany opened in 1955 as a result of a partnership between BASF and the British division of Shell.[33] Bayer and British Petroleum teamed up in 1958 to build a plant in Dormagen. French planners made extensive efforts to integrate the only major IG Farben–derived plant in their zone (BASF Ludwigshafen, including the Oppau plant nearby) into their nation's economy.[34] The international business community had every opportunity to learn about German techniques during the occupation years. Despite that, the widespread interest in know-how gave the Germans a real bargaining position. What remained was a question of just how enforceable such know-how agreements would be in courts around the world, whatever the business community's demand.

The Legal World of Know-How

As know-how licensing rose in popularity, it fell to business lawyers and legal scholars to hash out what exactly the vague term meant in the precise language of contracts and to draw upon any parallels they could find in common law precedents. The sharp spike in business interest in know-how, and thus know-how contracts, led in turn to a very similar spike in legal discussion of scientific and technological know-how. The percentage of articles in law review journals using the term "technical know-how" reflects the growth of interest in the concept.[35]

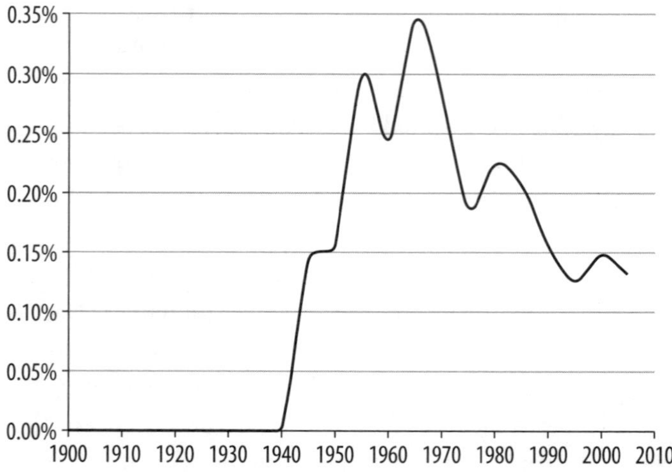

Figure 7.2. Percentage of law review articles that contain the term "technical know-how" in the HeinOnline law review journal collection.

One of the most fascinating parts of the rise of know-how is the way in which this almost necessarily vague term interacted with the precise and concrete world of business law. For businesses, this generality was useful. A 1949 article in the trade journal *Electric Engineering* pushed firms to consider the new trend of know-how licensing, describing them as "provid[ing] the foreign manufacturer with reliable 'de-bugged' designs in return for . . . payment," as well as "another less tangible but equally desirable feature from the standpoint of the foreign licensee . . . a technical listening post within the United States where a large fraction of the world's technical research is being carried out today."[36] Those licensing their know-how to others received extra revenue and an opportunity to explore foreign markets without the commitment of wholesale expansion.

For lawyers, vague language was inherently a problem. A contract was of little value if courts would refuse to enforce it when one side did not follow through. If one side claimed that the other had not fully transferred the know-how behind their process, how were courts to judge? Many components of modern-day intellectual property law did not yet exist in the 1940s, including common usage of the term "intellectual property" itself. Could lawyers find enough common law precedents dealing with skilled technicians, employee mobility, trade secrets (another term without clear definition at the time), technology transfer, and other related topics to build an intellectual property right to know-how, such that it would be the equal of trademarks or copyright? Many business lawyers were optimistic that a property right to know-how could be defined and enforced, but it required tackling many very difficult conceptual and technical legal problems.

A cottage industry emerged around this topic as attorneys interested in licensing, labor law, and technology transfer began writing article after article on the protection of unpatentable ideas. These articles had titles such as "A Question of Property Rights: The Government and Industrial Know-How," "Licensing Know-How, Patents, and Trademarks Abroad," and " 'Know-How' Licensing and Capital Gains" and were written on subjects such as the tax implications of know-how agreements, whether know-how agreements faced the same restrictions as patent agreements when it came to antimonopoly enforcement, and many other topics.[37] Almost every article, though, struggled by starting afresh on defining know-how.

The definition and redefinition of technical know-how in the law is beyond the scope of this chapter, but a few examples might help show how this was both an emerging, important issue to business lawyers and one for which they had little solution. Kingman Brewster, professor at Harvard Law School (and later president of Yale University), defined know-how in his influential 1958 book *Antitrust and American Business Abroad*:

> For convenience, we shall call all unpatented information "know-how." Know-how, however, may mean several different things. It may consist of designs, formulas, and processes which could be patented but for reasons of nondisclosure were not. Or it may be a highly personalized skill accumulated over years of experience which cannot be communicated or taught except in person. It may be objectively recorded information embodied in manuals which represent the continuing accumulation of solutions to production "bugs" and problems. Or it may be the provision of a personal advisory service, managerial or technical, expert but not unique.[38]

Many articles began with a disclaimer about the impossibility of defining the term:

> The term "know-how" is not susceptible to exact definition. In the broadest sense, it may consist of inventions, processes, formulae, or designs which are either unpatented or unpatentable; it may be evidenced by some form of physical matter, such as blue-prints, specifications, or drawings; it almost invariably includes trade secrets; and it may involve accumulated technical experience

and skills which can best, or perhaps only, be communicated through the medium of personal services. It can be seen that know-how as a general descriptive term comprehends a variety of forms and natures.[39]

Some definitions veered almost into poetry:

> Know-how, in contrast [to patents], is more likely than not to be an amorphous, ill-defined glob of technology that has no clear time limits and no clear geographic limits. Its subject matter is not only vaguely defined; it is not even publicly defined. As likely as not it is ever changing, like a stream of water flowing through a fish pond, as old know-how becomes public property and new know-how is added to the batch. . . . Know-how, in terms of content and legal status, is like a cloud in the sky that forms, dissolves, forms again, shapes and reshapes as the atmospheric conditions change.[40]

Yet lawyers continued to define and redefine know-how over and over for decades to come, because it remained a key issue for their clients.

All of this discussion comes only from the American case, yet know-how became an issue in legal systems around the world: common law and civil law, some lacking the peculiar American distrust of monopolies, some with greater fear of foreign industrial hegemony, some more worried about technology being denied, and others more worried about technology being stolen. How to define *el know-how technico*, ноу-хау, 노하우, ノーハウ, *das Know-how*, and transliterations in other languages became an international legal issue as the term rose in popularity. A full international, comparative history of trade secrecy, industrial espionage, and technology transfer in this period awaits its historian, but it is clear that know-how became a priority in international business and law in the decades following the war, at least in part through the wide exposure of businessmen and policymakers to the problem of capturing German know-how.

Industrial Technology Transfer in the Diplomacy of the Early Cold War

Governments, too, became far more interested and involved in controlling transfers of industrial technology during the early Cold War. Promoting the spread of science and technology became a key part of Cold War diplomacy on both sides of the Iron Curtain, and even across it. This was also a broader phenomenon than the intellectual reparations programs, but both drove and were influenced by these programs. One of the justifications American and

British policymakers gave for taking German science was that reports would be available to the whole world and would thereby generate goodwill while standing in contrast with Soviet secrecy. As FIAT wrapped up, this same line of thinking—and some of the same infrastructure and personnel—turned toward building alliances by sharing a different bounty of industrial science and technology: "our great American production Know-how."[41]

Sharing American know-how became an important plank of the Marshall Plan, a slew of related programs aimed at improving worker productivity in allied Western European nations, and economic development programs in the developing world. Allies with stronger economies meant allies with stronger defensive capabilities if war broke out and potentially less turmoil that communists could use to seize power. A similar logic appealed to Soviet decision-makers in preparing for what they saw as an inevitable confrontation with the capitalist world. Throughout all of this planning, lessons from the occupation of Germany—especially, but certainly not limited to, the importance of know-how—imbued this newly enriched link between diplomacy and industrial science.

Preaching Productivity: Technical Aid and the Construction of Postwar Europe

In the early postwar years, the Board of Trade, headed by Sir Richard Stafford Cripps, worried deeply about reversing what they saw as a decline in British industry and, more generally, in British world power. Britain also owed enormous war debts, primarily to the United States. These concerns had driven the Darwin Panel and BIOS, and in the postwar years they drove a number of related foreign policy and industrial policy decisions.

By July 1948, Cripps—now chancellor of the Exchequer—approached Paul Hoffman, Marshall Plan administrator, about a different solution for strengthening British exports: acquiring American technology and management know-how. The result was the Anglo-American Council on Productivity (AACP), which quickly became the most prominent Marshall Plan program in Britain.[42] Much like with BIOS, while the initiative came from the United Kingdom, US agencies quickly took a leading role (to the annoyance of some British counterparts) in this endeavor.

The AACP mirrored the later stages of BIOS operations. With a budget of close to a million British pounds, most of which was provided by the United States, the program sent 138 teams of "managers, workers, and specialists— over 900 people in total" on extended trips to tour American firms.[43] In order to really take in American know-how, the program planned for each visit to last four to six weeks. Like the BIOS teams, the AACP teams were designed to balance large and small firms across industries as well as to represent management, labor representatives, engineers, and technical workers. The teams wrote summary reports about their findings. These reports were reproduced and sold (and sold well—more than 600,000 copies in total) as advertisement for the new gospel of productivity.[44]

The program, which lasted through June 1952 before reorganizing under a different program title, cost about 3 million USD. The US Economic Cooperation Administration contributed about 2.1 million USD of that. While the AACP began as an exchange program, it quickly became one focused on getting British productivity up to American standards, reacting to 1940s economic statistics showing that British workers in the 1930s were about half as productive as American workers.[45]

The economic impact of the AACP is difficult to assess. Based on the general tone of wrap-up reports by participants, media attention at the time, and historians' assessments, gains were likely modest and indirect. Many of the British teams visiting the United States were dominated by upper management types, despite early planning to include on-the-ground workers, and these managers tended to put more stock in psychological issues (e.g., British workers not being sufficiently "production minded" and the United States having a "climate of productivity") than in specific technologies or organizational methods.[46] As they saw it, American technology was extremely effective at leveraging the United States' abundant natural resources, large domestic market, and workforce—but would it translate to British industry? Change would require not only money but upending long-standing British industrial culture, such as a standard of strictly hierarchical management composed of nontechnical personnel. Some British managers took the lesson to heart and enacted meaningful, useful changes, but many were content to go on as before. (Curiously, if this was also a problem of integrating German techniques, it never arose in any of the sources I have consulted—possibly because the dominance of BIOS teams by engineers and on-the-ground technicians prevented this managerial conservatism and fatalism from taking hold.)

The American side of the AACP, led by cochairman Philip Reed of General Electric, was frustrated by the failure of these sustained visits to produce concrete gains. Some began to suspect that British industrial leaders were uninterested in actual change. One such American planner commented in 1952 that "during the past two years I have talked with more than 200 British businessmen . . . [and] officers of the Board of Trade, the Monopolies Commission, AACP, officers of Trade Associations and other industrialists. No one has uttered the words so dear to the heart of an Evangelist, 'What can I do to be saved?' "[47]

The Americans certainly wanted to measure and quantify the impact of their programs, but British businesses rarely cooperated, as they were worried about the loss of trade secrets. Trade journals from individual industries include mixed reports. A 1955 review of the bronze and brass casting industry written by the British Productivity Council (the successor to the AACP and thus in a position to benefit from a rosy report) concluded that "the adoption of new techniques has not greatly extended in the last two years, although many more firms have come to realise their value. While the more enlightened firms have continued a policy of purchasing new plants or reconsidering their production programs, others have been content to look on with an 'it costs too much' attitude or have ignored the recommendations entirely."[48] On the other hand, the journal *Business* reported in 1955 about one manager from this industry who "had completely transformed his firm's operations after taking part in an AACP team visit."[49]

Whatever the direct impact, it seems likely that, similar to BIOS in Germany, the AACP and its successor agencies played a role in building connections among businesses by providing a much clearer picture of each business' needs and capabilities. In 1958, US firms surveyed about the AACP and similar productivity tours from other countries reported that these programs drove know-how licensing: "the technical missions visiting in the United States had increased inquiries and eventually [our] licensing agreements."[50] British planners, in turn, hoped that better communication would inspire long-term investment and partnership, if only through clearing up misconceptions. The *Economist* criticized the AACP's "rather facile assumptions" that improving British productivity could be as easy as sending teams to America but applauded that "the Americans on the Council could now disabuse their countrymen of such widespread myths as that Britain's difficulties were due largely to sloth."[51]

Even in nations more skeptical of American diplomatic leadership (e.g., France), similar technical aid programs were attractive. French communists and their left-wing allies saw American capitalism and consumerism as real problems, and right-wing industrialists disliked America's evangelism for trust-busting. Still, American industrial and scientific might was unquestionable, and France's needs were great. Jean Monnet, a businessman and influential diplomat who was later crucial to the formation of the European Union, spent the duration of the war traveling to Britain and America, where he saw first-hand the incredible potential of large-scale, standardized production. Monnet hoped to harness some of these lessons in addressing one of France's most pressing needs: coal.

Few issues were as important economically and for national security as the coal-producing regions of the Ruhr and Saar, located just across the prewar border in now-occupied Germany. This area had been a tinderbox that helped ignite two world wars. Its combination of coal reserves, geography, and pre-existing industrial development gave it the potential for tremendously productive coal and steel industries, both of which were vital for rebuilding and, if necessary, rearming. Gaining as much control of these regions as possible was a top priority for French leaders in postwar diplomacy. This goal drove discussion around the formation of a French zone of occupation, French obstruction in the Council of Foreign Ministers in Berlin, and French economic and military planning in general.[52] One result was that the French zone of occupation, which encompassed this area, had more of these "low-tech" coal and steel facilities and fewer "high-tech" chemicals and other fields than did the American zone, further playing into French attitudes toward exploiting German science and technology (discussed in chapter 3).

Monnet and like-minded industrialists turned to American technical aid to expand and modernize the French steel industry. They did so despite reservations about whether American methods would necessarily work in France, since—as French leaders had weighed even more heavily in the exploitation of German technology—they felt that technology was fundamentally embedded in the society around it. Adoption of American production methods was slow.[53] Still, French industry was dogged, and American aid programs reported better relations with French than with British industrialists: "the technical-exchange aspects of the program were quickly and enthusiastically received by the French [and they developed] the largest and most varied productivity program in Europe."[54] Historian Matthias Kipping argues that this success was due to "the entrepreneurial initiative of a few key individuals, most of whom were active businessmen rather than 'technocrats' or government officials."[55] This was the key difference between the French wartime exploitation programs and postwar partnerships with American technical aid programs: differing optimism about the possibility of technology transfer.

Sharing (or jointly creating) science and technology was an important part of the push toward a united Western Europe. American technical aid and support for rebuilding Western European science was only one part of that story, though it is an important one.[56] Bilateral and multilateral agreements sprang up even among countries that had been on opposing sides in two total wars in recent memory. In 1950, France formally agreed to form the supranational European Coal and Steel Community, creating a common market and encouraging cooperation in managing these products. The European Coal and Steel Community, in turn, served as a model for the 1957 Treaty of Rome. As a result of the treaty, the European Economic Community and the European Atomic Energy Community (dubbed Euratom) were established. Though Euratom ultimately failed to shift the locus of nuclear research from national programs to a centralized and shared European research program, it was another site for building a sense of Europe as more than a collection of fully independent, sovereign states. It also helped stitch together networks of elite experts across national lines, a purpose for which science has been a powerful tool throughout the twentieth century.[57]

Science and technology being seen as "apolitical" made them an ideal area in which governments could partner even with those with whom they had recently been at war. The inability of even medium-sized governments to fund "Big Science" by themselves served as another justification for bringing Europe together. The complex story of how rival nations became the European Union cannot be told without substantial attention to scientific-technical collaborations within Europe, as well as the external support of American and Soviet aid.

The Soviet Bloc

The Soviet Union stands apart from the other nations discussed here in some fundamental ways, making direct comparisons difficult. While the United States, the United Kingdom, and France each sought to use science and technology to build up alliances and influence in Western Europe, the Soviet Union retained firm control over Eastern Europe by way of the more explicit threat (and, as in East Germany in 1953, a reality) of military intervention. As a result, while it is worth discussing how science and technology played into the relationship between these vassal states and the Soviet Union, there is at minimum a different inflection on what it means to say the Soviets sought to build "diplomatic goodwill" akin to American technical aid programs.

East Germany turned to the Soviet Union for technical assistance a number of times throughout the 1950s. As the world turned from coal-based chemistry to petroleum-based chemistry, East German industry needed help in acquiring both physical goods (above all, the petroleum itself) and the technologies involved. In 1958, East Germany and the USSR signed an agreement to build a petroleum chemistry–based production plant in Schwedt, near the Polish border.[58] In return for cash payment, the USSR agreed to provide blueprints and refining equipment, and to exchange expertise—both sending Soviet technicians to help get the plant running and training German workers in Soviet facilities. Given the cash payment in return, this transaction was not even at the level of philanthropy of the AACP. Still, this and other joint projects had a real opportunity cost for the Soviet Union, relative to selling this oil and equipment abroad. Building up ties within the socialist world, and building up an interdependent Soviet bloc economy, was a priority, and technical aid played an important role.

Soviet technology was not always an easy fit, even within the centralized economies of Eastern Europe. In 1959, a key East German chemical industry official outlined a number of real problems in technology transfer from the Soviet Union: primarily, the Soviet tendency to build enormous-scale facilities meant real challenges scaling down. Despite these challenges, historian Raymond Stokes found that "although the deployment of Soviet technology at Schwedt required enormous effort to adapt it to conditions in the GDR, Soviet assistance was absolutely essential to construction and operation of the plant."[59] The Soviet Union was not as committed in other fields, even within the chemical industry, and ignored requests for technical aid in many other areas.[60] The end result was that Eastern European countries continued to seek technology transfer agreements with the West throughout the Cold War, to whatever extent this was possible between Soviet disapproval and Western technology export controls. Still, technology transfer was part of stitching together the Soviet bloc economies and building ties among these nations.

Around the same time the United States started its formalized technical assistance programs for Western Europe, the Soviet Union began its own mission to build up Chinese science and industry. After the Communist Party of China won its bloody civil war against its nationalist rivals, talks began for Soviet aid to the new communist state in early 1949. Funds and equipment were, of course, a major part of this package. Like the Marshall Plan, though, a series of treaties in the early 1950s between the USSR and China expanded its scope to emphasize technical assistance, combining physical equipment with the expertise of Soviet specialist teams sent to China.[61] By 1952, representatives from these countries reached an agreement to jointly build fifty important projects. The next year, two new agreements expanded the industries targeted, expanding the total to about 150 enterprises. The Soviet Union agreed to send experts to help build power plants and other infrastructure, as well as factories to produce a wide range of products: heavy machinery, machine tools, tanks, tractors, beatings, instruments, and so forth.

Soviet technical assistance to China is especially interesting because of the complex relationship between Mao's regime, Soviet leadership, and the technicians and scientists whose expertise (and frequently, whose Western training) made them a threatening, separate source of authority. By the late 1950s, these technical assistance programs had transferred equipment, but often with insufficient know-how to use it. This problem was exacerbated by the scarcity of technically trained Chinese technical specialists, as so many had been purged. Mao expressed his frustration with Soviet aid in 1958: "As to heavy industry—its design, construction, and installation—all could not be done by the Chinese. Not having experience or experts in China, we had to imitate foreign countries, and even with imitation, we could not always duplicate the technology. Furthermore, we had to rely on Soviet experience and experts to overcome the bourgeois ideology of earlier Chinese experts. Additionally, although most Soviet designs adopted for use in China were sound, some were flawed, but were copied without thinking."[62] Of course, we should not take these sorts of complaints entirely at face value, since these aid programs were tied into a much longer story of tension between the Soviet Union and China over whether and how far the latter should defer to the former's leadership, as well as to doctrinal differences. This Sino-Soviet split peaked in the mid-1950s, though Soviet technical aid for the Chinese nuclear weapons program continued in an effort to retain influence and build goodwill.

Despite the polarization of the Cold War and technology export controls discussed later in the chapter, there was a real market for East-West technology trade. The know-how craze spread to the Soviet Union as well, not least because they sought to license technology from the West. The term was transliterated into Russian as ноу-хау by the 1960s, though the best translation was still considered to be секреты производства.[63] American chemical firms in the late 1950s were hesitant to license out know-how. Industry leaders met with military officials at the Fifth National Military-Industrial Conference in Chicago in 1959, where they collectively decided that trade with the USSR should include only finished chemical goods, not the know-how to produce them.[64] The Soviet Union, by this point led by Nikita Khrushchev, found opportunities elsewhere. Britain's Imperial Chemical Industries, one of the largest chemical manufacturers in the country, was one of many suppliers of know-how to Soviet enterprises. In 1964, the arrangement was going well enough that they even agreed to expand their exports from just know-how to selling both know-how and physical plants.[65]

The extent to which Western companies were willing to trade know-how with the Soviet Union also varied over time, as 1940s-era suspicion of the Soviets as ignorant gave way to post-Sputnik perceptions of the Soviets as scientifically sophisticated. As one British licensing expert commented in his 1967 advice to would-be Western licensors: "In view of the considerable achievements of Soviet technology a clause should be negotiated [into know-how contracts] that the Soviet party will communicate to the licensor all improvements made by it or its sublicensees."[66]

Know-How in Global Economic Development

By the 1950s, Cold War tensions and paranoia led to a new US policy: communism must be contained around the world, and any individual country "going red" was a threat to American national security. In the extreme, this meant proxy wars such as Korea and Vietnam were justified as crucial battles between capitalism and communism. On the softer side, the extensive international propaganda campaigns of the Cold War led the United States and Soviet Union to expand both science and technology and foreign aid.[67] As the United States, the Soviet Union, and other nations took new interest in the economic development of what they dubbed "Third World" nations, the know-how craze spread beyond advanced economies seeking to protect their own innovations.

The question of how to help a poor nation "modernize" its economy was an unsolved and difficult one.[68] Traditional imperialism did not even try to turn colonies and protectorates into highly developed nations, so there were few concrete examples from which to draw. President Harry Truman's inaugural address in 1949 described one theory for how it might happen: "The United States is pre-eminent among nations in the development of industrial and scientific techniques. The material resources which we can afford to use for assistance of other peoples are limited. But our imponderable resources in technical knowledge are constantly growing and are inexhaustible. . . . I believe that we should make available to peace-loving peoples the benefits of our store of technical knowledge in order to help them realize their aspirations for a better life. And, in cooperation with other nations, we should foster capital investment in areas needing development."[69] The International Technical Cooperation Act of 1949 (aka the Point Four Program, named for being the fourth foreign policy objective in this inaugural address) enacted this mission. "Point Four," Truman explained, "means making our scientific advances and technical know-how available for the improvement and growth of underdeveloped areas. Point Four means technical missionaries at work."[70]

Point Four was a compromise with budget hawks who were very much worried about "the material resources which we can afford to use for assistance." In testimony before Congress about the act, undersecretary of state (and future NASA administrator) James Webb argued that "the exchange of persons and ideas does not require heavy expenditures, as do supply programs. Moreover, technical knowledge and skills can be shared without loss to those who now possess them. In fact, those who do share them will themselves learn much through seeing how they can be adapted to different conditions and through learning about the skills developed in other parts of the world."[71]

Webb reminded Congress that "the idea of exchanging knowledge and skills is not new. We have been participating with other nationals in such programs for years." His specific examples were from academic science: the Interdepartmental Committee on Scientific and Cultural Cooperation, the Fulbright Act, and related programs. Since the subject matter here was applied, industrial, and especially agricultural production techniques, he might have added US Department of Agriculture programs that worked with Latin American farmers throughout prior decades, distributing seeds and information. He also might have mentioned FIAT, wartime interallied exchanges, and the postwar productivity missions.

As historian David Ekbladh has emphasized, for policymakers in the 1950s (guided by social scientists' recommendations), "capital was not enough. The problem was most underdeveloped nations just did not have the capacity to employ the aid. . . . Technical assistance was necessary to tutor people in modern techniques [and] . . . this required 'sustained participation of private as well as public authorities.'"[72] With Congress' approval on June 5, 1950, the State Department set to work by way of the newly founded Technical Cooperation Administration to implement this goal. The program's budget for 1950–1951 was 25 million USD—ten times what the United States contributed to the AACP but aimed now at the entire developing world.

Point Four combined concerns about diplomatic positioning with national security concerns about releasing dual-use technologies, or releasing any technologies at all to the Soviet bloc. In public, technical aid program officials emphasized that agreements would only be signed with nations who they felt could be trusted to mutually engage and cooperate. Behind the scenes, even the secrecy-minded Truman administration feared technology leaking out less than they sometimes proclaimed to a McCarthy-era public. A talking points memo sent around to Truman administration representatives in April 1949 addressed whether the program would be available to Soviet bloc countries: "Off the record, we do not feel that the provision to the Iron Curtain countries of certain types of technical experts, under the auspices of the United Nations, would necessarily be more to the advantage of the Iron Curtain countries than to ours. Few information or propaganda programs are more effective than such technical experts."[73]

The United Nations had its own technical aid program based on the circulation of technical experts with know-how. The first such mission started in December 1946: finding experts to help Greece improve its agriculture and fishery sectors. In December 1948, the General Assembly gave the secretary-general authority to "arrange for the organization of international teams consisting of experts" more broadly, to be combined with training programs of various kinds for experts in these "under-developed" areas.[74] In the early 1950s, UN technical aid programs received between 20 million and 45 million USD per year as they expanded. In comparison, the US Technical Cooperation Administration in the State Department received 100 million USD in 1952 and about 140 million USD in 1953. The Mutual Security Agency, a successor to the Marshall Plan's Economic Cooperation Administration, provided an additional 200 million USD in economic and technical aid just in Southeast Asia and the Pacific areas in 1953.[75] The American programs, then, were an order of magnitude better funded.

Both the American and UN programs demonstrated a remarkable shift in thinking from the initial planning around the intellectual reparations programs. "Know-how" was now the watchword, in part because of a belief that it could work wonders (without necessarily losing anything back home) and in part because it was seen as a cheaper alternative to investing capital and granting low- or no-interest loans to impoverished nations. This new philosophy went so far that it began to receive criticism by the 1960s. As one pair of commentators argued in 1962: "The original Point Four Program (and for that matter the UN Technical Assistance Program) publicized the notion that the overseas job could be done by pushing the business of exporting 'know-how' and assuming that, once the bearer of this magic phrase was on his way to the country of operations, he would miraculously catalyze the poor, benighted, unknowing, indigenous population."[76] Thinking about technology might have evolved from seeing it as a shiny bauble that could be easily picked up, wrapped in a report, and copied, but in at least some cases, this know-how itself became "a commodity exportable, similar to, yet cheaper than machinery."[77]

Technology, and the language used to talk about it, depends on the culture in which it is set, and the allure of know-how was not necessarily quite the same from the perspective of those receiving this aid. As one French legal expert who specialized in licensing agreements with "Third World" nations described it: "The term 'technology,' as used by third world countries, goes considerably beyond its traditional concept as understood in western countries, and covers in addition to scientific and industrial know-how, operational and managerial know-how, such as how to organize and operate industrial, agricultural, touristic and other types of large projects. In fact, a growing number of transactions have recently involved exclusively this type of know-how."[78]

Developing nations faced complex policy questions in regards to technology transfer and intellectual property of all kinds, whatever their exact understanding of know-how. The United States was eager to partner with developing nations, but—despite (or perhaps because of) being a country that built its own industrial might in large part on ignoring other nations' intellectual property rights—it was also eager to protect its businesses by pushing American legal norms around the world.[79] This put these developing nations in something of a bind, as their own firms might well benefit by selectively enforcing patent rights.

America's history of business-driven imperialism in Latin America was another shadow looming over developing nations' reception of American technological missionaries. Huge companies such as United Fruit used local resources and employed unskilled labor for much of the nineteenth and twentieth centuries but kept any sophisticated production techniques secret and shipped profits away. Would the many businesses offering technology transfer agreements, as part of Point Four or independent of it, be similarly exploitative? For those worried that patent licenses were just setting up local businesses to be eternally dependent on foreign firms, know-how licenses seemed like a real solution. At the end of a know-how license, a firm could theoretically be competitive in world markets, even if the licensor cut off all ties (or if the industry were nationalized). For this and other reasons, American commentators felt that by the 1960s, "know-how has . . . assumed an aura of fascination in newly developing countries which see in it a mystical factor which may resolve or bridge over the difficult initial steps of technical and economic development."[80]

By 1979, researchers estimated that "approximately 90% of all licensed technology received [in developing nations] is of the know-how variety."[81] Similarly, the former director general of the Mexican Registry of Technology Transfer estimated in 1976 that "more than 75% of technological licensing agreements do not involve patents and . . . fall within the category of know-how licensing contracts. Extensive research conducted at the international level demonstrates that in the future the trend will be to rely more on know-how licensing and to gradually reduce the use of patents as a main object in the contract (with exceptions in certain fields, such as pharmaceuticals)."[82] More research is needed to see whether and how this know-how focus worked out in practice in these developing countries (if most 1960s and 1970s development projects are a guide, probably not that well), but the *interest* and the *effort* are themselves a demonstration of changing international thinking about technology transfer, science, and society.

Denying Technology: Hoarding Scientists and Encouraging Brain Drain

While spreading technology to allies took on new importance for governments, so, too, did denying it to rivals. This tension between jealously guarding the secrets of technologies and recruiting "technical missionaries" to share them existed even within the exploitation agencies in Germany. Agencies such as FIAT and BIOS were tasked with preparing reports about German technology for the benefit of all, and were even asked to facilitate independent investigators from non-Allied nations. Yet they were also to hoard as many skilled scientists and technicians as possible, as many patent documents and research results as possible, as many rocket parts and machine tools and wind tunnels as possible, and above all as much nuclear material and knowledge as possible in order to deny those technologies to the Soviet Union— and sometimes even to deny them to *any* other Allied nation. As the wartime alliance soured into Cold War enmity during the late 1940s and early 1950s, this mission of technological denial expanded from investigations in Germany to a major part of international diplomacy.

Denying Technology as a Priority in Occupied Germany

Some programs were more oriented toward denial than others. Operation Paperclip and Operation Surgeon (a British program aimed specifically at aeronautics, unlike the more general Darwin Panel), for example, were very much aimed at denial first and industrial benefit second. When US forces first invaded lands that were to become part of the Soviet zone, they ordered about 1,800 scientists and technicians (and three times as many family members and staff) to move west toward the soon-to-be US zone. A particularly egregious example of this is a set of about 450 scientists and engineers evacuated from the area around Jena in July 1945. The group included world-class specialists in optics and electronics from Carl Zeiss in Jena, Schott and Genossen, and the universities of Jena and Breslau.

Almost a year later, they and their families sat in internment camps, never even having been interrogated. The scientists were "unanimous in their opinion that their scientific and technical talents ha[d] been completely wasted," and only after they repeatedly prompted FIAT were they assigned to writing reports on German science and industry as part of a make-work "FIAT Technical Institute."[83] The only real goal in many cases was keeping skilled personnel from Soviet hands. The US Department of Commerce only gained Truman's approval for Operation Paperclip by emphasizing this justification of denying skilled manpower to unfriendly nations, and this reasoning helps explain why so many scientists who were of marginal value for US firms or had questionable "denazification" ended up on US Air Force bases.[84] The issue was less what America would gain than what the Soviets (presumed to be behind German standards) would not gain.

Also, too, was at least as much about denying nuclear technology to the Soviet Union as it was about discovering how far along Nazi nuclear research had gotten. The mission, led by Colonel Boris Pash and scientific advisor Samuel Goudsmit, first evacuated nuclear and biological weapon–related files and personnel from any Nazi counterattacks. Later, when key scientists and files were heard to be located in areas about to be captured by the French army and become the French zone of occupation, they organized a mission to deny the French access to nuclear research.[85] They raced behind enemy lines to dismantle and evacuate nuclear research tools and moved dozens of scientists west as well.

However, even programs relatively more focused on acquiring industrial technology, such as FIAT (US) or BIOS, had denial among their top priorities. When FIAT (US) decided that a certain scientist (Dr. Ronald Richter, a Czech physicist working in Berlin on "changing the surface structure of metals to improve catalytic effects of metals, separation of isotopes and light weight storage batteries") was of particular value, the official recommendation they sent to US forces in Europe was for the "Exploitation and Denial of German Scientist."[86] The British also saw this as a priority. In January 1947, nearly every agency involved in policy toward Germany (the Board of Trade, Foreign Office, Control Commission for Germany, British Element [CCG/BE], Colonial Office, and Ministries of Defence, Treasury, Admiralty, and Supply) met to discuss government-sponsored industrial research in the British zone of occupied Germany. Their highest priorities were both helping British industry and implementing "the policy of denial of German scientists to the Russians."[87]

Overlapping with the later stages of these intellectual reparations programs was the next generation of technological denial initiatives. Among them were the CoCOM, a "gentleman's agreement" to deny technology to the Soviet bloc, and American initiatives to stop the spread of nuclear technology to new countries. Neither Paperclip nor CoCOM was the first major governmental attempt to restrict the spread of technology to other countries. In the context of the Second World War and early Cold War, however, science and technology took on new importance for governments around the world. Spreading technology became a key tool of diplomacy, and diplomacy a key tool of stopping the spread of technology.

CoCOM: A Gentleman's Agreement to Wage Economic War

As early as 1945, American planners began working with Europeans to prevent the Soviet Union from buying militarily useful products or licensing the know-how behind them. The Export Control Act of 1949, the first peacetime export control measure in US history, handed the executive branch enormous power to institute export control "for national security purposes." The Department of Commerce began drafting licensing requirements for companies seeking to trade with the Soviet bloc, while the Department of State set out to work with allied (mostly Western European) nations on ensuring that they, too, would deny these technologies. At first, attempts to work out export controls through bilateral deals ran up against a problem of trust: why prevent your own industry from making money by licensing technologies if another country would do it anyway? By 1950, however, leaders from the United States, the United Kingdom, France, and other nations settled on an informal but effective way of negotiating and monitoring what they would and would not all agree to embargo: the CoCOM.

Deciding what technologies to embargo—and, crucially, what restricting technology exports even meant—was its own headache. The United States sought an extremely broad definition of military-use technologies, and its 1950 proposed list of most important (Class 1A) technologies reads like a list of

FIAT targets: "specialized machine tools (40 items), petroleum equipment (15 items), chemicals and chemical equipment (31 items), precision scientific and electronic equipment (42 items), and certain nonferrous metals (12 items)."[88] France and Britain, while willing to embargo technologies they saw as legitimate national security risks, were far more selective and more concerned with supporting their export industries (which, after all, comprised a substantially higher percentage of their economies than did exports for the United States). After putting together their own proposed lists, they merged them in 1949 into an Anglo-French proposal. Eventually, CoCOM settled on a compromise list, including the most urgent American priorities, and keeping discussion and debate open on other items.

Export controls, unlike the first phrase of exploitation programs in Germany, set out from the start to target unclassified, unpatented technical data and "know-how." New regulations issued by the Department of Commerce in 1949 as part of America's investment in CoCOM defined technical data as "any professional, scientific, or technical information, including any model, design, photograph, photographic negative, document, or commodity, containing a plan, specification, or descriptive or technical information of any kind which can be used or adapted for use in connection with any process, synthesis, or operation in the production, manufacture, reconstruction, servicing, repair, or use of any commodity." While extremely broad, this would seem to focus on material goods. Yet the "technical data" definition was expanded to include "advanced developments, technology, information, and 'know-how,' including prototypes and special installations, and those items listed in Proclamation 2776 [arms, ammunition, and implements of war, essentially] which do not have a security classification, whenever they have significance to the common security and defense." Department of Commerce officials conceded that defining all relevant types of technical data would be difficult, but they would be focusing on advanced, security-related technologies and "know-how" necessary to install and operate them. As a result, Commerce officials also wanted to know about foreign firms wanting to visit US plants producing such technologies.[89]

The decision to sacrifice economically in order to slow Soviet acquisition of technology was not an obvious one. One strand of thinking among American policymakers in the mid-1940s suggested "containment by integration"—that is, drawing the Soviet Union into the international economy and diplomatic scene, and hoping this would lead to economic leverage. As President Truman put it, the hope was that the USSR "was susceptible to pressure, especially economic pressure, which could be used to control, discipline, and punish it."[90] This thinking got swept aside by growing Cold War tension abroad and anti-communist hysteria at home, but did not go away entirely. Faced with the threat to British exports that CoCOM represented, British prime minister Winston Churchill commented in 1954 that "the more trade there is between Great Britain and Soviet Russia and the satellites, the better still will be the chances of our living together in increasing comfort."[91]

Technology export controls were an expensive and incomplete solution to a problem, and were supported in no small part by anti-communist fervor, but the problem of Soviet industrial espionage was a real one. The brief opening of archives after the collapse of the Soviet Union, combined with rare instances in which American intelligence agencies broke Soviet ciphers, has given us a few glimpses into Soviet espionage in America, and it was extensive.[92] The Amtorg Trading Corporation, or Amtorg, was the official trade representative for the Soviet Union in the United States since the 1920s and served more or less openly as a front for acquiring technology ever since. During the Great Depression and then wartime alliance, many companies were more than willing to allow Soviet representatives to inspect production methods before placing orders. This provided opportunities for Soviet representatives to learn about manufacturing techniques even for products that were not available for export. Meanwhile, KGB operatives recruited scientists and engineers to pass additional information along for pay, or (as was often the case, especially before the horrors of Stalinist society became clear to the West) because the informant believed in the hope of a worker-owned society.[93] By the end of the Cold War, this state-sponsored industrial espionage became so widespread that the CIA created a special unit to combat it.[94]

Nuclear Know-How: Denial and Diplomacy

Nuclear technology was at the heart of many policy debates about state intervention in the spread of technology during the Cold War. As such, it ties together many of the themes discussed previously in this chapter: the know-how phenomenon, technology as diplomatic leverage, and nonproliferation as a key foreign policy goal. A full accounting of the complex history of nuclear technology across the Cold War world is well beyond the scope of this book, but it is worth at least broaching some of the ways that this nuclear story fits into and diverges from these broader trends.[95]

Policy discussions about nuclear technology, for example, were very much shaped by the know-how phenomenon yet also show how ideology and politics could intervene in how policymakers handled technology controls. In the late 1940s, American and British policymakers, alongside many Manhattan Project alumni, argued explicitly that the only real monopoly America had over nuclear technology was industrial and scientific know-how. By the late 1940s to 1950s, however, opportunistic politicians exploited fears (and realities) of atomic espionage, the first Soviet atomic bomb, and communist infiltration in general, and turned the conversation to protecting a "secret formula" behind the bomb. Political maneuvering and propaganda were sufficient to change the conversation away from know-how, at least for this specific technology.

In the months and years following the dropping of atomic bombs on Japan, President Truman was one among many voices assuring the public that know-how was the key to America's nuclear monopoly. In October 1945, just a month after the war's official end, Truman gave a speech arguing this point, which the *New York Times* reported under the title "US Will Not Share Atom Bomb Secret, President Asserts: Calls Industrial 'Know How' the Most Important Factor, Not Scientific Knowledge."[96] Truman argued that even if allies such as Britain and Canada possessed the knowledge, only the United States had the industrial capacity and resources needed for the bomb. A similar logic lay behind the release of the Smyth Report (technically titled *A General Account of the Development of Methods of Using Atomic Energy for Military Purposes under the Auspices of the United States Government, 1940–1945*), which laid out the principles behind the atomic bomb and was released to the public just days after the bombings of Hiroshima and Nagasaki. Protecting this information seemed less important than informing the public, especially since so much of it was in the public domain already. The really important information was the know-how.

British leaders made the point even more clearly. As Winston Churchill explained to Parliament in November 1945, the only real "secret"

> is the practical production methods, which they have developed at enormous expense and on a gigantic scale. This [nuclear proliferation] would not be an affair of scientists or diplomats sending over formulas. To be effective, any such disclosure would have to take the form of a considerable number of Soviet specialists, engineers and scientists visiting the United States arsenals, for that is what the manufacturing centers of the atomic bomb really are. They would have to visit them and would have to dwell there so they could have it all explained to them and the officials would then return to their own country with

all the information they had obtained and with any further improvements which might have occurred to them.[97]

Clement Attlee, prime minister of the United Kingdom from 1945 to 1951, further instructed Parliament about the nature of technology transfer when speaking about a forthcoming atomic energy bill in October 1946:

> The production of atomic energy involves very complicated processes. It is really a major industrial effort, and until we can get international control, what is sometimes called the industrial "know-how" must be kept under control. When I was in America the declaration made by the President of the United States, the Prime Minister of Canada and myself laid down this policy: until we can get the introduction of effective and forcible safeguards, and we all hope that international arrangements will make strict secrecy unnecessary, while we can meanwhile encourage the dissemination of basic scientific information, there must be power to prevent the dissemination of information as to what is called the "know-how."[98]

Historian David Kaiser argues that from 1945 to 1948, this message began to get through to journalists and the broader public. "Most coverage focused on raw materials and industrial infrastructure as the keys to producing atomic weapons, emphasizing non-textual 'know-how' rather than textual 'knowledge' or 'information.' Beginning late in 1948 and accelerating through the mid-1950s, the weight of discussion among politicians and journalists shifted, focusing instead on textual and theoretical 'information' as the essential 'secret' of the atomic bomb, rather than experimental skill or industrial capacity. Many now claimed that specific, esoteric formulas—the x's and y's of theoretical physics—contained the true secrets of the atomic bomb."[99]

This shift came in large part, Kaiser argues, through the engineering of the House Un-American Activities Committee (HUAC). As HUAC investigated espionage within the Manhattan Project, they leaked information that—taken out of context—implied that scientists were careless with crucial, codified documents that threatened the nuclear secret.[100] *Time* magazine, among others, bought into this hype around a "Hot Formula," and an ongoing press offensive by HUAC and like-minded McCarthyites drove this message home. Reports of actual atomic spies such as Klaus Fuchs gave a face to these accusations. The first Soviet atomic test happened years to decades before most American intelligence experts have predicted, augmenting the general fear of Soviet espionage. By the 1950s, however, many experts closely involved with science and technology might insist that it was nuclear know-how that mattered, know-how largely disappeared from the political discussion and newspaper reporting. Instead, the American people feared the disloyal scientist willing to smuggle formulas and blueprints across the Iron Curtain.

This policy focus on explicit knowledge rather than know-how served the United States very poorly in combatting the threat of technological espionage throughout the Cold War. Concern over leaked documents led to an enormous, opaque system of classified documents and security clearances, segmenting the world of academic, open science from the much larger and better-funded world of classified science.[101] Meanwhile, top-level scientists and engineers contributing to American military research were exiled, ensuring that their know-how was spread in the most effective way. A particularly egregious example is Tsien Hsue-shen, a US Air Force colonel and expert on jet propulsion technology who was born in China. After graduating from MIT and Caltech, he helped study the V-2 missiles brought back from Germany during the occupation and made major contributions to American rocketry. During the height of anti-communist hysteria in the late 1940s, Tsien was accused of communist leanings (based on almost no evidence) and was deported. Once in China, he helped develop a rocketry program there and contributed far more than any microfilm ever could have contained, even if Tsien had been inclined to send such reports.[102]

Atoms for Peace and the Diplomacy of Sharing Nuclear Technology

While much of the story of nuclear technology fits more closely with the "denying technology" theme, by the 1950s American policymakers were willing to share nuclear technology, too, for diplomatic gain—albeit only civilian nuclear power technology, and only after the Soviet Union had already developed an atomic bomb of their own anyway. In December 1953, President Dwight D. Eisenhower announced his Atoms for Peace program at the United Nations. Through Atoms for Peace, the United States would provide samples of radioactive isotopes to researchers around the world for medical and scientific purposes, and even assist in constructing nuclear reactors for research and power production.[103]

Atoms for Peace was hardly pure philanthropy. Eisenhower saw it as a tool of "psychological warfare" to win the "struggle for the minds and wills of men."[104] Making a show of sharing civilian nuclear technology served a number of priorities. It would distract from the buildup of nuclear weapons that were part of Eisenhower's "New Look" strategy. Insofar as the Soviet Union was willing to take part by showing off their own nuclear power technology, it would provide invaluable intelligence about Soviet capabilities. Ideally, it would even pressure the Soviet Union into spending funds on nuclear power technologies *instead of* military technology. Most of all, it would build up America's image in the world. As the National Security Council wrote in 1955, "Such a program will strengthen American world leadership and disprove the Communists' propaganda charges that the US is concerned solely with the destructive uses of the atom. Atomic energy, which has become the foremost symbol of man's inventive capacities, can also become the symbol of a strong but peaceful and purposeful America."[105]

The program was a major shift from the Truman administration's strategy of denying nuclear technology to others at all costs. The Atomic Energy Act of 1946 had classified all nuclear technology, placing it under the control of the civilian Atomic Energy Commission and implementing strict restrictions on licensing out or otherwise releasing atomic information. The Atomic Energy Act of 1954, in comparison, significantly loosened restrictions on exporting nuclear technology, so long as the partner nation accepted certain safeguards. For example, a country might allow a US contractor (e.g., General Electric or Westinghouse) to build a nuclear power plant or research pile in their territory and accept enriched uranium fuel the United States provided, so long as the spent fuel was returned to the United States afterward. The newly formed International Atomic Energy Agency served as a venue to develop multilateral safeguards and standards. Meanwhile, the United States signed bilateral agreements with more than twenty countries within the first two years of the program and almost double that many by 1961. Participant countries included European countries (Belgium, Greece, Portugal, Spain) as well as developing nations (Argentina and Brazil, among others).

Providing the information necessary to run and sometimes build civilian nuclear projects while denying the experience, skills, and materials to work toward nuclear weapons was a delicate balancing act, and it was not always successful. Israel, India, and Pakistan are among the nations who acquired at least some of their nuclear power technology with American technical aid, then used that know-how in the pursuit of nuclear weapons.[106] Still, even this risk—what was considered enormous from a national security perspective— was apparently worthwhile in the chase of building tighter alliances against feared Soviet aggression.

For the Soviet Union, too, sharing nuclear technology was an important tool for alliance-building. Soviet leaders were reluctant to share nuclear power technology for the same reasons as many American officials. In addition to the economic cost of this kind of assistance, as historian Sonja Schmid put it, "Soviet leaders feared that exporting civilian nuclear technology to Eastern Europe would entail the spread of sensitive, dual-use materials and know-how."[107] Despite that, Soviet propaganda often highlighted American militarism, including the keynote of its use of nuclear weapons in Japan, and ceding the imagery of the "peaceful atom" to the United States was unacceptable. As historian Paul Josephson writes: "The peaceful atom showed that a nation whose citizens had been illiterate and agrarian less than forty years earlier, had become a leading scientific and industrial power. The achievements of science and technology, with nuclear energy at its summit, were symbols of the legitimacy of the regime both to Soviet citizens and to citizens of the world. The peaceful atom also allowed the USSR to score points with the conquered countries of Eastern Europe . . . each of whom had a nuclear program based on Soviet isotopes, technology, and training programs and, in part, its largesse."[108]

Soviet scientists and technicians eagerly participated in UN-sponsored Atoms for Peace conferences, the most prominent of which was the 1955 conference in Geneva on peaceful uses of nuclear technology. While the United States and United Kingdom sent the largest contingents, the Soviet Union sent seventy-eight representatives. Scientists from both sides of the Iron Curtain were enthusiastic about exchanging what information they could and building personal relationships.[109] Like the United States, the Soviet Union amended its law specifically for the 1955 conference in order to allow the declassification and publication of information on nuclear power technology.

Within Eastern Europe, the Council for Mutual Economic Assistance, established January 25, 1949, provided an institutional framework for Soviet technical aid. Science and technology were central to Soviet programs to build peaceful ties with these Eastern European allies (to accompany the threat of military intervention). As Schmid argues, "nuclear cooperation was perceived as part of a shared modernization effort" between the Soviet Union and Eastern Europe, and this modernization through science was in turn seen as apolitical.[110] The details of this technical assistance in nuclear power were unclear and evolved over time but included turnkey reactors, designs, technical information, nuclear fuel, and perhaps most importantly, expertise. By the 1970s, more than three thousand Eastern European scientists and technicians were given opportunities to train in Soviet establishments, and more than one thousand Soviet experts spent time in Eastern Europe working on nuclear projects.

There were limits, of course, and often the same ones that the United States imposed on Atoms for Peace deals. Recipients of nuclear fuel had to return the spent fuel so that it could not go toward nuclear weapons research. The USSR offered only one type of light-water reactor. The Soviet nuclear power industry was itself in relative infancy in the 1950s (as was the American industry), and there was not immediately an overabundance of expertise to share. Still, these bilateral (and, later, multilateral) agreements led to some Eastern European nations having a substantial portion of their electrical power coming from nuclear plants by the late Cold War. In Eastern Europe, German specialists who had returned to the country after Operation Osoaviakhim were important in developing that industry, as were other local experts trained in Soviet techniques.[111]

Again, there is an enormous amount of nuance being passed over here, and those wishing to know more about the geopolitics of nuclear proliferation and nuclear technology should consult the many sources cited throughout this book. The main point here is to emphasize the importance Cold War policymakers placed on the spread of science and technology. In the 1940s, the key to containing technologies seemed to be secrecy, restricting access to material resources, and depending on the difficulty of developing (let alone "stealing") know-how. By the 1950s, the reality of a Soviet nuclear bomb and the desirability of gaining diplomatic standing through sharing nuclear technology (in addition to the other policy goals of Atoms for Peace) made technical aid too powerful of a tool to ignore, even in this sector.

Conclusion

One of the lasting consequences of the exploitation of German science was a realization, shared by businessmen and policymakers around the world, that sustained, on-the-ground contact is the most effective way to transfer technology. That might seem almost obvious, yet as the planning behind the exploitation of German science shows, it has not always been so clear. The correction toward pursuing German "know-how" happened at the start of a powerful, largely forgotten trend in twentieth-century business, law, and government toward planning around know-how. The Cold War era political interest in technology transfer spanned ideologies and industries, and governments around the world became far more interested in controlling the flow of technology across their borders.

For private industry, this way of thinking about technology transfer had some counterintuitive effects. Planning around taking German technology often pictured it as a zero-sum game. By the 1950s, however, American industry concluded that they often stood to *benefit* from other nations licensing their know-how, due to American advantages in infrastructure, technical personnel, and organization. Not only would exchanging engineers with a foreign firm bring in licensing revenue, but it could provide insight into overseas markets without the risk or expense of foreign direct investment. Meanwhile, technology transfer agreements were generally a two-way street, and American firms were in a strong position to seize upon any clever innovations made by their partners. In West Germany, the economic consequences of technical exploitation were, at absolute worst, insufficiently damaging to prevent the "economic miracle" (*Wirtschaftswunder*) of the 1950s. West German firms came to many of the same conclusions as the Americans who had been on the other side of the interrogation table—they stood to gain by leveraging government-sponsored investigative trips to figure out who knows what, and then follow up with private, mutually beneficial licensing agreements. America's lead was far from absolute, and German firms still had a great deal of invaluable know-how that greatly interested American (and other nations') firms.

On a national scale, breaking from thinking of technology as a zero-sum game had far-reaching effects. As the proliferation of technical aid programs demonstrates, governments became more willing to share science and technology for strategic ends. Historians have been debating what was one conventional wisdom, "that the transfer of US technology and managerial know-how lay at the heart of the extraordinary economic growth experienced by Western Europe and Japan during the 'golden-age' of the long post-war boom," but it likely played at least a meaningful role.[112] The effects on the developing world are likely more important still, though untangling the technology transfer dimensions of development planning is a major project awaiting historians' careful attention.

Conclusion

As I have attempted to show in this book, there was far more to Allied intellectual reparations efforts than a few "Nazi scientists" being hired for rocket research. The idea that it was worth investing time, effort, and political and diplomatic capital to acquire German science and technology appealed to each of the major Allied powers (and many other nations not discussed in this book but which took part as well), despite their enormous differences. It appealed to military leaders, politicians, and civilian bureaucrats covering a wide range of governmental departments in each nation. It appealed to businessmen and representatives of trade groups across industries as diverse as chemicals, wood pulping, fishing, toy making, audio recording, synthetic rubber, metallurgy, automobiles, prefabricated housing, aeronautics, textiles, kiln manufacturing, scientific equipment, and machine tools. Even ignoring other programs, FIAT alone published almost fourteen hundred reports on specific industries and technologies by the end of 1947, and that does not take into account the many investigative groups that reportedly never submitted final reports in their fields.[1] Thousands of German scientists, engineers, and skilled workers found new employment (whether voluntarily or not) in service of the conquering powers. The entire world of German patent filing, contents of scientific libraries, and records of firms' industrial research were available for international competition to search.

The scale and ambition of these programs are worth emphasizing, even if the amount of science and technology they truly transferred to the Allied nations seems to be much less than planners had hoped (or opponents feared).

Despite (or perhaps in some small part because) of these investigations, West German industry thrived in the 1950s, in the famous "economic miracle" (*Wirtschaftswunder*). This book is not a history of German economic history, nor even one focused on the entirety of postwar reparations politics, but there seems to be little evidence that at least the Western Allies' programs significantly hurt West Germany.[2]

Anyone who wants to argue that the intellectual reparations were an enormous gain for the United States, for example, can find plenty of evidence to support that claim. In some industries, there very clearly were substantial, important gains. This is especially the case in military technologies such as jet engines, submarines, and especially rocketry. Magnetic tape for audio recording crossed from Germany to America through FIAT investigations.[3] Many companies expressed wholehearted enthusiasm for the information they gained while investigating Germany, and their claims must be taken just as seriously as the detractors. Among the many glowing reports in the archives of TIIC, FIAT, BIOS, and other investigative agencies are reports like one that states, "one report alone saved him $140,000," or Goodyear reporting that another "would save them at least $20,000 in research."[4] A journalist for the *Daily Express* estimated the value for the United Kingdom at £100,000,000 as early as October 1945, though he admitted that estimations varied so wildly that he could as easily guess £1,000,000,000 without anyone being able to disprove it.[5]

Still, in most industries, the full process of technology transfer (including not only collecting information but then also adopting, adapting, and utilizing it in a new context) did not occur.[6] In this book, I have mostly emphasized the evidence pointing in the other direction. That is, though I make no attempt to put a dollar amount on these intellectual reparations, I think in most cases their direct economic impact is exaggerated. There is plenty of evidence to find here, too. The archives of FIAT, BIOS, and the Comité de coordination scientifique de la defense nationale (FIAT [France]) have plenty of expressions of cordial thanks and cooperation, but they also are full of frustration and disappointment. Bureaucratic duplication and infighting prevented efficient or speedy travel, study, processing, and publication. Many industries seemed unwilling to contribute investigators or translators. Though many businesses and foreign governments bought copies of FIAT and BIOS reports, the reports never reached as many people as the programs' leadership had hoped. Very often, investigators expressed disappointment that upon closer inspection, German technology was not actually going to be particularly useful for their company or industry.

The greater importance of these efforts was probably not the economic value accrued but rather that the attempt was made at all: the lessons it taught those involved with technology transfer, the diplomatic consequences of pursuing these ends so aggressively, and the consequences of releasing this massive backlog of scientific and technical information for the intelligence communities and scientific communication institutions (including library networks). Without simply repeating the conclusions to each chapter, it is worth laying out more precisely exactly why and how the postwar investigations of German technology matter without anchoring them in a dollar figure. One way of phrasing this importance on a national level is simply by asking what each country gained and lost.

National Assessments

Despite what American and British intelligence agencies perceived, France likely came out ahead. To these erstwhile allies, French policy was underhanded, duplicitous, and manipulative, trying to undermine German science so that they could have their agents "steal" the best Germans from other occupation zones. Some of that likely happened, and France certainly did bring hundreds of V-2-related technicians into temporary camps and laboratories to help the French military acquire a basis in that technology. In general, though, the team that was sponsored by the Centre national de la recherche scientifique (and coordinated with other departments) focused instead on what they saw as the practical, feasible approach: long-term integration between French and German research establishments.

As economic historian Alan Milward has argued, French pursuits of a European economic community centered on Franco-German ties (once the United Kingdom made clear that it was uninterested, at least) "were the true determinants of a more lasting Western European settlement."[7] In his study of the German chemical industry in the early postwar years, Raymond Stokes made this connection directly: "Throughout the period 1945 to 1948, German and French technical personnel played a pivotal role in Franco-German relations," and "French actions in Germany were bound up with the origins of postwar Franco-German *rapprochement*, which has been a key determinant of European political and economic development since 1945."[8]

To be sure, there was a diplomatic cost to France not fulfilling allies' requests for faster cooperation in copying German science via report-writing and extracted scientists. French inattention to quickly duplicating Anglo-American investigative teams' reports spawned conspiracy theories about the French having found ultravaluable scientific secrets, contempt for French lethargy, and general frustration. Further, the *stagiaires* plan was a complete failure in terms of the trainees acting in their originally planned capacity as intelligence agents in German labs. Still, the efforts to integrate Franco-German research institutes and rebuild German science had long-term beneficial effects, not least in binding those deeply suspicious, hostile nations together.

In at least one prominent vein of British history, the BIOS investigations and reparations policy in general was a "missed opportunity" to prevent British "industrial decline."[9] As historian David Edgerton has shown, this trope of British industrial decline has a long history and does not track particularly well with reality in the postwar decades, but it remains an important cultural belief. Contrary to this idea, I argue that while the United Kingdom likely never received an enormous boost to its export industries as planners in the government had hoped, British firms probably gathered some modest but real value from German industry. Earlier planning, centered on Anglo-American cooperation in investigations and report-writing, seems to have been something of a failure in terms of actually transferring technology. As businesses and trade groups complained about the impossibility of acquiring "know-how" in this way, though, BIOS and the Board of Trade responded. Longer trips, with British engineers staying in German firms to work closely with technicians there, seem to have generated more encouraging reports. Only larger firms who could afford to send over technical investigators could benefit from this sort of approach, which was a betrayal of original planning that emphasized being fair to entire industries. Still, at least some British firms seem to have gotten a "bite out of the apple" rather than industry getting just a "smell all around."

The United Kingdom did have some substantial advantages over any of the other major Allied powers. Its economic position, level of industrial technology, and political institutions were in much better shape than either France or the Soviet Union. It was also a much shorter trip to Germany for British teams than it was for Americans, which meant many more could afford to go. Exact figures are difficult to come by, but John Gimbel cites 1,400 total American investigators through the end of May 1947, whereas Carl Glatt lists 1,300 British investigators just from July to August 1945 and another 2,800 from August 1945 through February 1946.[10]

Conversely, developing the goodwill of scientists and elites was a priority for the British occupation authorities, and ongoing BIOS efforts could not have been beneficial in that regard. It seems unlikely that the United Kingdom had a substantial, foreseeable "missed opportunity" in terms of intellectual reparations, but there were costs paid for whatever knowledge was gained, not all of them financial.

For the United States, the economic stakes were probably lower than for anyone else. Wartime state investment had vastly expanded industrial capacity. The United States had lost a substantially smaller percentage of its population than other combatants. The destruction in Europe had eliminated most rivals for world markets. Possibly the greatest challenge facing American industry was converting its excess industry to productive use and finding jobs for the returning soldiers.[11] If all FIAT's hopes had been realized, maybe it could have been in an even more powerful position, but it was hardly make-or-break for American industry. German scientists made important contributions to both military and civilian aerospace technology, but even many of the hundreds of scientists brought over by the infamous Operation Paperclip sat in barracks for months or years as the air force struggled to find businesses interested in hiring them.

Both Soviet and American planners emphasized the importance of "scientist denial" programs. In practice, though, however important know-how and specialized knowledge undoubtedly were, both nations were willing to invest enough resources to overcome whatever deficit was left by the potential German scientists. This is particularly noticeable in the Soviet atomic bomb program. American intelligence agencies debated how long it would take the Soviet Union to develop an atomic bomb, and estimates ranged from years to decades (or, in some estimates, never).[12] Nearly all were shocked by the first test explosion in August 1949, in large part because they had vastly underestimated the capabilities of Soviet scientists. Had the Western intelligence agencies hired every German chemist, physicist, and technician, the historical consensus seems to be that it would only have added a few months to the Soviet program, still far before American estimates.

The greater consequences for America were in terms of lessons learned about how to communicate science and technology. That might seem like a minor and vague gain, but scientific communication and technology transfer became very important in both diplomacy and business in a quickly globalizing economy. When America tried to rebuild Western Europe's economies —and then developing nations outside of Europe—science and technology, and, in particular, America's "know-how," were central to that planning. This was true in the development of the Anglo-American Council on Productivity and the deployment of on-the-ground technical experts in development efforts in Venezuela and elsewhere.[13] Private firms, too, began to question how they could benefit from transferring American technology through know-how, and this led to changes in international intellectual property law during the Cold War period. There were certainly important causes aside from FIAT for these major changes in business, science, and diplomacy, but for those thousands of businesses and dozens of government agencies across many nations, FIAT was an important reason to give serious, systematic thought to these issues.

The USSR went its own way in taking German technology. At low-level talks, the Allied nations considered allowing intellectual reparations teams to cross the line between the Soviet zone and the Anglo-American Bizone, and a few such trips even took place under close supervision of the Soviet or American hosts. Still, there was nothing like the interallied coordination that convinced the French to set up FIAT (France). If one main story about American investigations centers on gathering too much scientific information to process and most of it goes unused on microfilm reels in archives, Soviet industrial dismantling ended in a similar case of swallowing more equipment than it could digest. It is hard to tell the extent of this waste, but even if stories are exaggerated of unlabeled boxes of equipment rusting unassembled on rail platforms, the Soviet Union gained less than a more deliberate process might have offered. Of course, most of this disorganization was due to internal political rivalries, combined in the early years by soldiers and low-level cadres in the Red Army who refused any attempts by the high command to rein in looting, destruction, and violence. It can be hard to imagine a better-managed reparations process given those realities.

Soviet-owned firms represent one alternate mechanism for technology transfer, as did the Science and Technology Offices, and they likely provided Soviet industry with more effective access to German knowledge and skill. There, like in the occupation of the V-2 design and production facilities, Soviet planners showed an interest in keeping teams and environments together (as much as possible, given the conditions), akin to the prevailing thinking in France. Similarly, in Operation Osoaviakhim, the secret police and military government were so thorough in sweeping up full teams of researchers and their families that they even sometimes took maids, mistaking them for wives.[14]

Whatever problems the Soviets faced in learning from German technology, they never seemed to have considered merely writing reports, or hiring individuals when entire teams might be more productive.

Lessons and Legacies

It is always dangerous to extrapolate from the past to the future, but some lessons from the different nations' successes and failures seem especially relevant today. A key point is one that can seem academic but was a very concrete and difficult problem for businesses in the exploitation of German science in the following decades: science and technology are embedded in the societies around them, and moving them across national and cultural borders requires major, sustained effort. As I began to discuss in the previous chapter, judges, lawyers, and lawmakers involved in intellectual property law increasingly recognized this reality in the form of allowing widespread know-how licensing in the 1940s to 1970s. Today, trade secrets are one of the most commonly used intellectual property rights but also are rarely discussed in business schools or academic research institutions. Much more research into the history of trade secrecy and the roles it plays in modern-day society needs to be completed, but its importance is clear.

One area where industrial secrecy is especially policy-relevant is industrial espionage, especially in an age when governments are investing more into cyber security and are deathly afraid of foreign governments "stealing" technologies by pilfering data. As the FIAT investigations show, though, having all the written information about a technology you could want is no assurance that the recipient can reproduce it in a cost-effective, timely way. This hacking might still be a serious threat, but policy initiatives aimed at keeping America's technology edge might focus more on attracting and retaining foreign scientists and skilled workers in this country rather than forcing them out once student and temporary visas expire. Their know-how might be more valuable than what can be gleaned from any number of copied documents.

An interesting related question is what role information technology (or its lack) played in the outcome of FIAT and others. As I show from slightly different perspectives in chapters 5 and 6, part of the problem each nation encountered was that they faced a vast sea of information with tools inadequate to sorting, searching, or otherwise processing it. This was not a unique problem to scientific espionage—information overload was already reaching critical mass in academic science and in espionage agencies by the start of the war. Some early efforts at automating and machine sorting (e.g., Vannevar

Bush's prototype machines) seemed promising, and were yet another link between intelligence and science in this era, but they were far too rudimentary to contribute meaningfully.

To some degree, though, the planners involved in studying German science seem to have been caught up in a kind of technological enthusiasm common today. In their faith that microfilm would allow cheap, easy copying and transportation of vast amounts of information (which, relatively speaking, it certainly did), they let their overall ambitions run wild. Had they been more concerned with the limits of how much information could realistically be sorted, translated, copied, and put into practice back home, they might have considered much more limited efforts. As any law firm facing a "document dump" by opposing counsel can attest, dealing with numerous documents is no easy task today, even with the advances in information technology.

Businessmen and policymakers today should be wary of scare tactics about enemy firms or nations "stealing" technology, especially when limited to data breaches. There are many cases in history of technologies being reproduced (reverse engineered) using limited, written accounts or the finished product as a guide, but it is a slow, extremely difficult process. The bigger threat is the movement of people—but at the same time, employee mobility and immigration is the lifeblood of innovation. Study after study has found that inventors and other skilled workers moving from firm to firm benefits not only the new firm but also the old firm and the economy at large.[15]

There is no one-size-fits-all policy prescription that makes technology transfer easy to accomplish or to limit. The early stages of the exploitation of Germany show that a focus on written accounts will fail, whether it is copying German documents on microfilm or the 1990s business fad of "knowledge management," the goal of which was to capture workers' tacit knowledge in computerized databases. At the same time, know-how is not a cure-all. American technical aid programs for the developing world were not particularly successful, in part because they sometimes thought of know-how as a sort of mystical shortcut that obviated the need for major capital investment and other support. Disregard for the critical importance of human expertise led the United States to deport talented scientists for supposed communist ties during the McCarthy era, materially aiding (among others) the Chinese missile program. Conversely, faith in know-how as a wall that would shield America's nuclear monopoly led to overconfidence, as US policymakers and intelligence analysts drastically underestimated Soviet scientists' ability to reinvent nuclear weapons. Hard work, time, and some limited documentation to avoid going down mistaken paths were enough for these Soviet scientists to develop their own nuclear know-how.

Today, the pendulum has swung too far toward ignoring the vital importance of tacit knowledge, and once again we are heading toward a place where we fear that other countries will loot our proud technological excellence through digital copying. The lessons of the exploitation of German science should reassure us, at least somewhat, that "stealing" technology is not that easy. Our attention instead might turn toward immigration policies that bring bright minds into our universities and industries, then deport them after years of experience. From a protectionist standpoint, this outflow of expertise is a far greater threat. Perhaps, though, even this is shortsighted. The German experience shows us, too, that the personal relationships and business compatibility generated by this circulation of expertise can have powerful long-term benefits, even for those on the "losing" end of scientific intelligence missions.

Abbreviations

ANFF	Archives Nationale, Fontainebleau
BNA	British National Archives, Kew
NACP	The National Archives at College Park, Maryland
TNA: PRO	The National Archives: Public Records Office

Introduction

1. In this book, I use "Allied powers" to refer almost exclusively to the United States, United Kingdom, France, and Soviet Union. Other nations were part of the alliance, including China, Poland, Norway, and several others, and some even took part in intellectual reparations programs, but on a smaller scale. Similarly, I use "Axis powers" to refer to Germany, Italy, and Japan, though other nations were allied with or controlled by these nations during the war (including occupied "Vichy" France).

2. Commerce Secretary Henry Wallace used the term "intellectual reparations" to describe Operation Paperclip, but the term was popularized among historians by John Gimbel in his excellent book *Science, Technology, and Reparations*.

3. See the essay by Grattan-Guinness et al. in Parshall and Rice, *Mathematics Unbound*, 7–8; Siegmund-Schultze, *Mathematicians Fleeing from Nazi Germany*.

4. Interesting visualization and analysis of prize winners' nationalities over time can be found in Schmidhuber, "Evolution of National Nobel Prize Shares in the 20th Century."

5. On Germany's late nineteenth- and early twentieth-century reputation in science, see the essays in Grüttner, ed., *Gebrochene Wissenschaftskulturen*.

6. Alter, *The Reluctant Patron*, 96; Davis, "Friedrich Max Müller and the Migration of German Academics to Britain in the Nineteenth Century," 96.

7. Kevles, *The Physicists*, 14–45; Werner, *The Transatlantic World of Higher Education*.

8. Davis, "Friedrich Max Müller and the Migration of German Academics to Britain in the Nineteenth Century," 94.

9. Kirchberger, "Deutsche Naturwissenschaftler im Britischen Empire."

10. Von Gizycki, "Centre and Periphery in the International Scientific Community," 477.

11. On the chemical industry in the nineteenth and twentieth centuries, see Abelshauser, *German Industry and Global Enterprise*; Bellamy, *Profiting the Crown*; Berghahn, ed., *Quest for Economic Empire*; Lesch, ed., *The German Chemical Industry in the Twentieth Century*; Spitz, *Petrochemicals*; Steen, *The American Synthetic Organic Chemicals Industry*; Stokes, *Opting for Oil*.

12. Hartcup, *The War of Invention*; Kevles, *The Physicists*, chaps. 8–9; Rhees, "The Chemists' War"; Steen, "Technical Expertise and US Mobilization, 1917–18."

13. Steen, "Patents, Patriotism, and 'Skilled in the Art.'"

14. Berghahn, "Technology, Reparations, and the Export of Industrial Culture," 3.

15. Beissinger, *Scientific Management, Socialist Discipline, and Soviet Power*; Hughes, *American Genesis*, 249–94.

16. Berghahn, *The Americanisation of West German Industry*; Berghahn, "The Debate on 'Americanization' among Economic and Cultural Historians."

17. For just a sample of this extensive literature, see Giffard, *Making Jet Engines in World War II*; Heim, Sachse, and Walker, *The Kaiser Wilhelm Society under National Socialism*; Macrakis, *Surviving the Swastika*; Maier, *Forschung als Waffe*; Rammer, "Die Nazifizierung und Entnazifizierung der Physik an der Universität Göttingen"; Renneberg and Walker, *Science, Technology, and National Socialism*; Walker, *Nazi Science*.

18. On the emigration of German scientists to the West, and those who stayed, see especially Ash and Söllner, *Forced Migration and Scientific Change*; Gemelli, *The "Unacceptables"*; Krohn, *Intellectuals in Exile*; and Siegmund-Schultze, *Mathematicians Fleeing from Nazi Germany*.

19. Kevles, "'Into Hostile Political Camps.'"

20. On Nazism and modernity, see Biagioli, "Science, Modernity, and the 'Final Solution'"; Bauman, *Modernity and the Holocaust*; and Rieger, *Technology and the Culture of Modernity in Britain and Germany, 1890–1945*.

21. On these developments, see Biddle, *Dark Side of the Moon*; Manchester, *The Arms of Krupp, 1587–1968*; Neufeld, *The Rocket and the Reich*; and Neufeld, "The Nazi Aerospace Exodus."

22. Kirkpatrick, "The Clock Strikes Twelve."

23. "We'll Just Never Learn."

24. Glatt, "Reparations and the Transfer of Scientific and Industrial Technology from Germany," 893.

25. On von Braun, see Neufeld, *Von Braun*. On von Braun's team at the National Aeronautics and Space Administration (NASA) and their lives in America, see Laney, *German Rocketeers in the Heart of Dixie*.

26. Works on Operation Paperclip include Bar-Zohar, *The Hunt for German Scientists*; Bower, *The Paperclip Conspiracy*; Crim, *Our Germans*; Herrmann, *Project Paperclip*; Hunt, *Secret Agenda*; Jacobsen, *Operation Paperclip*; Kurowski, *Alliierte Jagd Auf Deutsche Wissenschafler*; Lasby, *Project Paperclip*; and Presas i Puig, "Reflections on a Peripheral Paperclip Project."

27. Gimbel, *Science, Technology, and Reparations*.

28. Gimbel, *Science, Technology, and Reparations*, 169–70.

29. Judt and Ciesla, eds., *Technology Transfer out of Germany after 1945*.

30. Stokes, "Assessing the Damages," 83.

31. Defrance, "La mission du CNRS en Allemagne (1945–1950)"; Defrance, *Les Alliés Occidentaux et Les Universités Allemandes, 1945–1949*; Defrance, Kißener, and Nordblom, eds., *Wege Der Verständigung Zwischen Deutschen und Franzosen Nach 1945*.

32. Villain, "France and the Peenemünde Legacy"; Villain, "L'apport des Scientifiques Allemandes aux Programmes de Recherche Relatifs aux Fusées et Avions à Réaction à Partir de 1945."

33. Examples include Chevassus-au-Louis, *Savants Sous l'Occupation*; Hecht, *The Radiance of France*; Hitchcock, *France Restored*; Kohlrausch and Trischler, *Building Europe on Expertise*; Milward, *The Reconstruction of Western Europe, 1945–51*; and Gillingham, *Coal, Steel, and the Rebirth of Europe, 1945–1955*.

34. Glatt, "Reparations and the Transfer of Scientific and Industrial Technology from Germany."

35. Farquharson, "Governed or Exploited?"; Uttley, "Operation 'Surgeon' and Britain's Post-War Exploitation of Nazi German Aeronautics."

36. Johnson, *Defence by Ministry*; Jones, *The Wizard War*; Maddrell, "British-American Scientific Intelligence Collaboration during the Occupation of Germany"; Maddrell, "Western Intelligence Gathering and the Division of German Science"; Maddrell, *Spying on Science*; Turner, "British Policy towards German Industry, 1945–49"; Wark, "British Intelligence on the German Air Force and Aircraft Industry, 1933–1939."

37. Albrecht, Heinemann-Grüder, and Wellmann, *Die Spezialisten*; Holloway, *Stalin and the Bomb*; Mick, *Forschen für Stalin*; Naimark, *The Russians in Germany*; Riehl and Seitz, *Stalin's Captive*; Sokolov, *Soviet Use of German Science and Technology, 1945–1946*.

38. Eckert, "Strategic Internationalism and the Transfer of Technical Knowledge"; Freeze, "Innovation and Technology Transfer during the Cold War"; Germuska, "Conflicts of Eastern and Western Technology Transfer"; Hagood, "Why Does Technology Transfer Fail?"; Harris, *Industrial Espionage and Technology Transfer*; Nilsson, "The Power of Technology"; Palmer, *Dictatorship of the Air*; Ramskjaer, "Users and Producers of Plastics in Post–World War II Norway"; Ryan, "The Role of National Culture in the Space-Based Technology Transfer Process"; Seely, "Historical Patterns in the Scholarship of Technology Transfer"; Seely, Klingner, and Klein, "'Push' and 'Pull' Factors in Technology Transfer"; Zhang, Zhang, and Yao, "Technology Transfer from the Soviet Union to the People's Republic of China."

39. Nash, "The Concept of 'Property' in Know-How as a Growing Area of Industrial Property," 289.

40. There is some reason to consider science separately from technology, but drawing lines between the two is extremely difficult, especially in the realm of industrial research. Most of those involved in these programs ignored the distinction and lumped together "science and technology." The war made clear that even abstract research could lead to nuclear bombs. Mobilizing science and scientists for developing the next generation of war-winning weapons became a priority for governments around the world. On the intertwining of science and military technology, see, for example, Dauchelle, "Le Réarmement Français Après la Seconde Guerre Mondiale"; Edgerton, *Warfare State*; Slaveski, *The Soviet Occupation of Germany*; and Stuart, *Creating the National Security State*.

41. "The Eternal Apprentice," *Time Magazine*, 8 November 1948.

Chapter 1 · American Exploitation Programs

1. An excellent, detailed discussion of this Paperclip program that brought over von Braun is in Crim, *Our Germans*.

2. Faragher, "Collecting German Industrial Information."

3. A main source for this section is Berghahn, "Technology, Reparations, and the Export of Industrial Culture." On American-German economic relations, see also Abelshauser, *The Dynamics of German Industry*; Berghahn, ed., *Quest for Economic Empire*; Lesch, ed., *The German Chemical Industry in the Twentieth Century*; and Steen, *The American Synthetic Organic Chemicals Industry*.

4. Quote drawn from Berghahn, *The Americanisation of West German Industry*, 29.

5. "Reports Germans Lead in Chemistry: Editor of Research Papers Says They Have Regained Pre-War Eminence," *New York Times*, 11 April 1930; Emil Ludwig, "Germany: Laboratory of the World," *New York Times*, 28 December 1930; "Industrial Uptrend is Noted in Germany," *New York Times*, 15 April 1935; "German Chemicals in Demand," *New York Times*, 6 September 1937; "U.S. Held Enriched by German Exiles: Flight of Chemists is Called Boon to Science Here," *New York Times*, 13 August 1939.

6. On these refugee intellectuals in general, see Fermi, *Illustrious Immigrants*; Fleming and Bailyn, *The Intellectual Migration*; Boyers, *The Legacy of the German Refugee Intellectuals*; Heilbut, *Exiled in Paradise*; Coser, *Refugee Scholars in America*; Jay, *Permanent Exiles*; McClay, "A Haunted Legacy"; Krohn, *Intellectuals in Exile*; Ash and Söllner, *Forced Migration and Scientific*

Change; Gemelli, *The "Unacceptables"*; Siegmund-Schultze, *Mathematicians Fleeing from Nazi Germany*.

7. Siegmund-Schultze, *Mathematicians Fleeing from Nazi Germany*.

8. Browne, "The Role of Refugees in the History of American Science."

9. Moser, Voena, and Waldinger, "German-Jewish Emigres and U.S. Invention."

10. Quoted in Macrakis, *Surviving the Swastika*, 14.

11. Macrakis, *Surviving the Swastika*, 14–21.

12. Studies detailing this timing include Berghahn, "Technology, Reparations, and the Export of Industrial Culture"; Morris, "Transatlantic Transfer of Buna S Synthetic Rubber Technology, 1932–45"; Stoltzenberg, "Scientist and Industrial Manager"; Wilkins, "German Chemical Firms in the United States from the Late 19th Century to Post–World War II."

13. Stokes, "Gravity and the Rainbow-Makers."

14. Tooze, *The Wages of Destruction*, xxii.

15. Tooze, *The Wages of Destruction*, 612.

16. Tooze, *The Wages of Destruction*, 661.

17. Goudsmit, *Alsos*; Rhodes, *The Making of the Atomic Bomb*; Walker, *Nazi Science*.

18. This quote is attributed to Secretary of State Henry Stimson, justifying in 1929 why he shut down the code-breaking agency developed during the First World War. Stimson and Bundy, *On Active Service in Peace and War*.

19. Gimbel, *Science, Technology, and Reparations*, 4.

20. For more detail on the creation of FIAT (France) and Anglo-American debates over how far they could trust France to cooperate in the exploitation efforts, see chapter 3. For more details about the establishment of FIAT, see History of the Field Information Agency Technical (FIAT) (8 May 1945–30 June 1946) 1st Installment (hereafter "History of FIAT"), in US Army Center of Military History, call number 8–3.2 AC v 1 c 2.

21. Gimbel, *Science, Technology, and Reparations*, 27.

22. Goudsmit, *Alsos*; Hart, "The ALSOS Mission, 1943–1945"; Pash, *The Alsos Mission*.

23. Gentile, "Advocacy or Assessment?"; Maclsaac, *Strategic Bombing in World War II*.

24. Krammer, "Technology Transfer as War Booty."

25. Bar-Zohar, *The Hunt for German Scientists*; Bower, *The Paperclip Conspiracy*; Gimbel, *Science, Technology, and Reparations*; Herrmann, *Project Paperclip*; Hunt, *Secret Agenda*; Kurowski, *Alliierte Jagd Auf Deutsche Wissenschafler*; Lasby, *Project Paperclip*; Peltzer, *Die Demontage Deutscher Naturwissenschaftlicher Intelligenz Nach Dem 2. Weltkrieg*; Presas i Puig, "Reflections on a Peripheral Paperclip Project."

26. "TIIC/C 8th Meeting," 26 April 1945. General Records, Department of Commerce, Record Group 40 (RG 40), box 31, National Archives at College Park, College Park, MD (NACP).

27. Gimbel, *Science, Technology, and Reparations*, 109–12.

28. Edward L. Bowles to Brigadier R. J. Maunsell, 25 June 1945. Records of the US Occupation Headquarters, World War II, Record Group 260 (RG 260.3.6), FIAT Administrative Records 1945–1947, box 17/1, folder 54, NACP.

29. "TIIC 34th Meeting," 24 August 1945. RG 40, Entry 75, box 30, NACP.

30. Lt. Col. Royal S. Copeland to Scientific Branch, 6800 FIAT, "Subj: Herman EHREN-SPECK, V-Weapon Research," 11 July 1946. RG 260.3.6, box 17/1, folder 25, NACP.

31. While not an ideal work in some ways, Hunt's *Secret Agenda* demonstrates the longevity of aspects of the Paperclip program.

32. "1947 Budget Estimates, Presented to Dept. of Commerce," October 1946. RG 40, Entry 75, box 58, NACP.

33. "Report on Work of Technical Information and Documents Unit," January 1950. The National Archives (TNA): Public Records Office (PRO), Board of Trade Record Group 211 (BT 211), folder 337.

34. Leonard, *More Than a Memoir*, 98–99.

35. Leonard, *More Than a Memoir*, 100.

36. Stent, *Nazis, Women, and Molecular Biology*.

37. Stent, *Nazis, Women, and Molecular Biology*, 40.

38. Stent, *Nazis, Women, and Molecular Biology*, 40–42.

39. Stent, *Nazis, Women, and Molecular Biology*, 43.

40. This section draws primarily upon Stewart, "The Office of Technical Services"; Gimbel, *Science, Technology, and Reparations*.

41. Stewart, "The Office of Technical Services," 64.

42. Green, "Last Call for Germany," *Federal Science Progress* 1 (February 1947): 24.

43. Gimbel, *Science, Technology, and Reparations*, 169–70.

44. Lehrer, *Wernher von Braun*. On von Braun generally, see, for example, Neufeld, *Von Braun*. On the von Braun team's experiences in America, see Laney, *German Rocketeers in the Heart of Dixie*.

45. For example, Jacobsen, *Operation Paperclip*; Hunt, *Secret Agenda*; Bower, *The Paperclip Conspiracy*; Simpson, *Blowback*. In terms of television, see, for example, Radice, "The Hunt for Nazi Scientists"; "The CIA and the Nazis."

46. For details on this, see especially Jacobsen, *Operation Paperclip*.

47. Some scientists were, indeed, eager to embrace Nazi ideology. Nobel Prize winners Johannes Stark and Phillip Lenard lent their reputations to "Deutsche Physik," a nationalistic brand of physics that rejected Einstein's relativity as "Jewish science." Other scientists resisted the regime. Many sought to avoid any political stance other than for independence of scientific institutions from direct state control. Joining the Nazi party was often necessary for any scientist in an administrative position to remain in control. Thus, almost all scientists in charge of research institutes were forced to join the party or flee Germany. Those who were only nominally members, with little or no further involvement, are worth separating from those who actively and eagerly supported the Nazi regime, however. There has been a great deal written about the role of science in the Nazi regime. See, for instance, Walker, *Nazi Science*, and Macrakis, *Surviving the Swastika*. For a more specific case study, see Heilbron, *The Dilemmas of an Upright Man*.

48. This discussion and following specific figures are from Gimbel, *Science, Technology, and Reparations*, 53.

49. Gimbel, *Science, Technology, and Reparations*, 55.

50. Neufeld, *Von Braun*, 89.

51. Giffard, *Making Jet Engines in World War II*; Neufeld, "The Nazi Aerospace Exodus."

52. Ciesla, "German High Velocity Aerodynamics and Their Significance for the US Air Force, 1945–52," in Judt and Ciesla, *Technology Transfer out of Germany after 1945*, 95.

53. Ciesla, "German High Velocity Aerodynamics and Their Significance for the US Air Force, 1945–52," in Judt and Ciesla, *Technology Transfer out of Germany after 1945*, 99–100.

54. Gimbel, *Science, Technology, and Reparations*, 150, citing "T. H. McConica to Ray L. Hicks, OTS, 11 December 1947, RG 330, JIOA General Correspondence, box 6, file Department of Commerce, NA."

55. Gimbel, *Science, Technology, and Reparations*, 224n26.

56. Krammer, "Technology Transfer as War Booty."

57. Krammer, "German Energy Secrets Recycled."

58. Morton, "'The Rusty Ribbon.'"

59. See, for example, US Department of Commerce, Office of Technical Services et al., "FIAT Reports."

60. Gimbel, *Science, Technology, and Reparations*, xi.

61. This term, utilized to describe German "inventiveness" contrasted to British "genius," was stated in *Hansard*, HL, 29 April 1948, Vol. 155, col. 568–90.

62. "Progress Report No. 1, Technical Industrial Intelligence Branch," 29 March 1946. RG 40, Entry 75, box 3, NACP.

63. "Press Release, Max A. Walz," undated. RG 260.3.6, box 17/3, folder 13, NACP.

64. "Press Release, PG Reynolds," undated. RG 260.3.6, box 17/3, folder 13, NACP.

65. "US Found Leading Reich in Engineering Technique," *Christian Science Monitor*, March 9, 1946.

66. "Survey of Productions Techniques Used in German Aircraft Industry," *Automotive and Aviation Industries*, March 15, 1945, 172.

67. Joseph Geschelin, "What Price Glory?," *Automotive and Aviation Industries*, May 15, 1946, 37.

68. Geschelin, "What Price Glory?," 37.

69. "German Machine Tools," *American Machinist*, February 14, 1946, 2.

70. "Doctors under Nazism," *Industrial Medicine*, September 1945, 28.

71. W. H. Reynolds to Robert Reiss, 5 September 1946. RG 40, Entry 75, box 3, NACP.

72. "German Fibers," *Chemical and Metallurgical Engineering*, November 1945, 292; "German Chlorine," *Chemical and Metallurgical Engineering*, October 1945, 104–6; "Acetylene: Industry in Wartime Germany," *Chemical and Metallurgical Engineering*, October 1945, 116–19; "FIAT," *Chemical and Engineering News*, August 25, 1946, 3297.

73. "Ueber Alles?" *Chemical and Metallurgical Engineering*, July 1945, 114–15.

74. J. C. Green to Assistant Chief of Staff, G-2, War Department, 14 May 1946. RG 40, Entry 75, box 12, NACP.

75. Lloyd Worden to Robert Reiss, 17 April 1947. RG 40, Entry 75, box 12, NACP.

76. It should be noted that a much stronger case for the importance of the investigations in Japan appears in Home and Low, "Postwar Scientific Missions to Japan." I cannot completely reconcile the differences, other than to note the same distinctions made elsewhere in this chapter between studying a technology and adopting it. Certainly, there were analogous efforts to the Alsos Mission to see how far Japan had pursued an atomic bomb, but those were short-lived. The efforts discussed in Home and Low seem to be focused mostly on military and economic intelligence, rather than on industrial investigations analogous to FIAT's scale or scope.

77. J. T. Kyd to G. S. Mansell, 28 February 1946. TNA: PRO BT 211, folder 41.

78. *Hansard*, HC, 12 May 1953, Vol. 182, col. 395–420.

79. J. Davidson Pratt, the Association of British Chemical Manufacturers, to Derek Wood, "Follow-Up Visits," 24 October 1946. TNA: PRO BT 211, folder 21.

80. "BIOS Exhibition: Notes on Telephone Conversation with Professor Linsted," 27 November 1946. TNA: PRO BT 211, folder 22.

81. For more on the specific struggles of the British efforts at scientific exploitation, see chapter 2.

82. Ralph Osborne to Chief of Staff, OMGUS, "Final Report of FIAT," 1 July 1947. RG 260.3.6, box 17/1, folder 50, NACP.

83. Bradley Dewey to Dr. Roger Adams, 19 December 1945. RG 260.3.6, box 17/14, folder 14, NACP.

84. T. J. Betis to Maj. Gen. Clayton Bissell, 3 June 1945. RG 260.3.6, box 17/1, folder 54, NACP.

85. "TIIC/C 8th Meeting," 26 April 1945. RG 40, Entry 75, box 31, NACP.

86. Joe Mayer, "Summary on OTS Exhibit at Food Exposition, Atlantic City, January 19–23," undated. RG 40, Entry 75, box 11, NACP.

87. Tell Berna to Capt. R. D. Syer, 31 July 1945. RG 40, Entry 75, box 12, NACP.

88. J. B. Quig to Howland Sargeant, 19 February 1946. RG 40, Entry 75, box 3, NACP.

89. S. D. Kirkpatrick to Robert Reiss, 26 November 1946. RG 40, Entry 75, box 3, NACP.

90. Albert J. Phillips to Robert Reiss, 4 September 1946. RG 40, Entry 75, box 3, NACP.

91. W. H. Reynolds to Robert Reiss, 5 September 1946. RG 40, Entry 75, box 3, NACP.

92. Wells, *Antitrust and the Formation of the Postwar World*.

93. The Eleanor Roosevelt Papers Project, "Interview with John C. Green, 22 August 1951, the Eleanor Roosevelt Show," https://scholarspace.library.gwu.edu/files/kp78gg381 (accessed 22 August 2018).

94. Major, "Chemistry in Postwar Europe."

95. Doel, "Scientists as Policymakers, Advisors, and Intelligence Agents."

96. Krige, *American Hegemony and the Postwar Reconstruction of Science in Europe*.

97. This war criminal dimension has most recently and thoroughly been examined in Jacobsen, *Operation Paperclip*, and Margolian, *Unauthorized Entry*.

98. "American, German, and Japanese Medical Reports: Publication List," 19 December 1946. RG 40, Entry 75, box 63, NACP.

Chapter 2 · *British Scientific Exploitation and the Allure of German Know-How*

1. Quote is from *Hansard*, HL, 29 April 1948, Vol. 155, col. 568–90, discussing exactly this issue of British inventions being neglected at home and capitalized on overseas, especially in Germany. On the importance of exports and anxiety over international competitiveness in business, see Cairncross, *The Price of War*, and Rollings, *British Business in the Formative Years of European Integration, 1945–1973*.

2. Hughes, "Notes on the Prospect Confronting Post-War British Patent Property," 736.

3. The history of ties between national security considerations and intellectual property law is an area in need of more research, but some quality work is emerging, such as Epstein, *Torpedo*, and Wellerstein, "Patenting the Bomb."

4. Most of this discussion draws from the memoirs of one of the British reparations negotiators after the Second World War, Alec Cairncross. See Cairncross, *The Price of War*.

5. The Board of Trade is very roughly analogous to the US Department of Commerce. It held a range of responsibilities related to commerce, industry, overseas trade, tariffs, patents, imports and exports, consumer protection, tourism, and statistics relating to these areas.

6. Derek Wood to Mr. Somervell, "Distribution of Industrial Information from Germany," 19 October 1946. TNA: PRO BT 211/24.

7. "Record of a Meeting to Discuss Further Actions in Connection with BIOS Group VII Investigation of German Wartime Patent Specifications," 17 April 1946. TNA: PRO BT 211/44; German Division, Board of Trade, "Facilities for Technical Investigations in Germany by Individual British Firms," 23 October 1946. TNA: PRO BT 211/21; D. J. Ezra to Derek Wood, 13 March 1947. TNA: PRO BT 211/24.

8. [Unknown] to V. B. Bennett, 16 September 1946. TNA: PRO BT 211/24; emphasis original.

9. W. G. Glennie, Deputy Regional Controller, "Bristol Exhibition of Industrial Intelligence Work in Germany," 28 December 1946. TNA: PRO BT 211/22.

10. Control Office for Germany and Austria, "Requests from British Firms for Facilities to Obtain Technical Information from Germany," 26 June 1946. TNA: PRO BT 211/21.

11. On libraries involved in publicity, see P. W. Bennett to the Undersecretary, Board of Trade, 31 July 1946; E. G. Davies to Miss E. F. Magg, Acting County Librarian, Wakefield, 17 April 1946. TNA: PRO BT 211/11; Gilbert LuDunn, Borough Librarian to the Assistant Secretary, German Department, Board of Trade, 26 June 1947. TNA: PRO BT 211/162. For other publicity arrangements, see Glennie, "Bristol Exhibition of Industrial Intelligence Work in Germany." "Darwin Panel Scheme: Draft Statement for Publicity," April 1948. TNA: PRO BT 211/476. "Minutes of the 6th Meeting of the Panel to Consider Employment of German Scientists, Specialists and Technicians in the United Kingdom," 30 January 1946. TNA: PRO BT 211/24.

12. "First Meeting of Panel to Consider the Employment of German Scientists, Specialists

and Technicians for Civil Industry in the United Kingdom," 3 December 1945. TNA: PRO BT 211/24.

13. C. H. Noton to Derek Wood, 17 July 1946. TNA: PRO BT 211/21. See also Economy and Industrial Planning Staff, "The German Clock and Watch Industry," January 1946. TNA: PRO BT 211/37.

14. German Division, Board of Trade, "Facilities for Technical Investigations in Germany by Individual British Firms," 23 October 1946. TNA: PRO BT 211/21.

15. Derek Wood to E. R. Wood, "Interrogation of German Scientists and Technicians in the UK," 18 October 1946. TNA: PRO BT 211/24.

16. Derek Wood to Mr. Somervell, "Distribution of Industrial Information from Germany," 19 October 1946. TNA: PRO BT 211/24.

17. Trevor Evans, "Smith and the Secrets of Schmidt," *Daily Express*, 9 October 1945. TNA: PRO BT 211/24; John Green to Derek Wood, 31 December 1946. TNA: PRO BT 211/24.

18. Derek Wood to K. Unwin, Esq., 10 October 1946. TNA: PRO BT 211/25.

19. Alexander King to H. L. Verry, 26 April 1946. TNA: PRO BT 211/17.

20. M. N. Salmond to Derek Wood, "Further Visits to IG Farben, Dormagen," 21 May 1946. TNA: PRO BT 211/21.

21. J. Davidson Pratt, the Association of British Chemical Manufacturers, to Derek Wood, 7 October 1946. TNA: PRO BT 211/21.

22. J. Davidson Pratt, the Association of British Chemical Manufacturers, to Derek Wood, "Follow-Up Visits," 24 October 1946. TNA: PRO BT 211/21; J. Davidson Pratt, the Association of British Chemical Manufacturers, to Derek Wood, 7 October 1946. TNA: PRO BT 211/21.

23. "Report on a Meeting of Technical Investigations in Germany by Individual Firms," July 1946. TNA: PRO BT 211/21.

24. "Technical Investigations in Germany by Individual Firms," July 1946. TNA: PRO BT 211/21.

25. "Report on a Meeting of Technical Investigations in Germany by Individual Firms," July 1946. TNA: PRO BT 211/21.

26. This account is from Longden, *T-Force*, 265–66. Longden wrote his account of T-Force activities on commission of T-Force veterans, using their files and recollections. While I cannot independently confirm this episode outside these veterans' memories, it seems highly plausible based on my own archival research in T-Force papers, and in any event speaks to what stuck out about the experience to these T-Force personnel: the shift from military priorities to catering toward specific civilian firms.

27. For the UK argument that it was "understood that the Americans are already attracting" such German personnel and that the United Kingdom faced "missing a valuable opportunity," see "Employment of German Scientists and Technicians in Civil Industry in the United Kingdom," 28 August 1945. TNA: PRO BT 211/46. For the similar American claim, see Gimbel, *Science, Technology, and Reparations*, 30–33.

28. Aldrich, "British Intelligence and the Anglo-American 'Special Relationship' during the Cold War"; Reynolds, "'A Special Relationship'?"

29. Neufeld, "The Nazi Aerospace Exodus."

30. PRO CAB 121/428, "The Recruitment of German Scientists and Technicians for Employment in the United Kingdom," Memorandum by the Control Office for Germany and Austria, DCOS (46) 217, 21 October 1946. As quoted in footnote 43 of Uttley, "Operation 'Surgeon' and Britain's Post-War Exploitation of Nazi German Aeronautics," 8.

31. "Employment of German Scientists and Technicians in Civil Industry in the United Kingdom," 28 August 1945. TNA: PRO BT 211/46; D. J. Ezra to Derek Wood, 13 March 1947. TNA: PRO BT 211/24.

32. "Minutes of the 3rd Meeting of the Darwin Panel to Consider Employment of German Scientists, Specialists and Technicians in the United Kingdom," 17 December 1945. TNA: PRO BT 211/24.

33. "Minutes of the 3rd Meeting of the Darwin Panel to Consider Employment of German Scientists, Specialists and Technicians in the United Kingdom."

34. "Minutes of the 3rd Meeting of the Darwin Panel to Consider Employment of German Scientists, Specialists and Technicians in the United Kingdom," and "Minutes of the Interdepartmental Meeting Held at the Board of Trade on December 20th . . ." 20 December 1945, both in TNA: PRO BT 211/24.

35. "Minutes of the Interdepartmental Meeting Held at the Board of Trade on December 20th . . ."

36. "First Meeting of Panel to Consider the Employment of German Scientists, Specialists and Technicians for Civil Industry in the United Kingdom," 3 December 1945. TNA: PRO BT 211/24.

37. "Report on a Meeting of Technical Investigations in Germany by Individual Firms," July 1946. TNA: PRO BT 211/21.

38. Derek Wood to S. T. Norman, Esq., British Commonwealth Scientific Office, 20 November 1946. TNA: PRO BT 211/24.

39. Derek Wood to S. T. Norman, Esq., British Commonwealth Scientific Office, 21 October 1946. TNA: PRO BT 211/24.

40. "Copy of Brief Handed to Major J. Day, Board of Trade, before Departure for Washington," March 1946. TNA: PRO BT 211/41; J. H. Wheatcroft, "Lists of Classified and Unclassified CIOS Reports, Amendment No. 3," 6 April 1946. TNA: PRO BT 211/13.

41. M. N. Salmond to Derek Wood, "Further Visits to IG Farben, Dormagen," 21 May 1946. TNA: PRO BT 211/21.

42. "Memorandum for Mr. Norman," September 1946. TNA: PRO BT 211/17.

43. W. G. Glennie, Deputy Regional Controller, "Bristol Exhibition of Industrial Intelligence Work in Germany," 28 December 1946. TNA: PRO BT 211/22. This six hundred investigators figure seems low for the US efforts, but British efforts involving substantially more investigators overall is highly plausible. Trips to Germany were cheaper and faster without transatlantic travel, so more companies could afford to send personnel, and the policy of including large and small firms across entire industries meant some groups sprawled in numbers.

44. J. H. Wheatcroft to G. Mansell, "Reports Awaiting Industrial Distribution," 6 September 1946. TNA: PRO BT 211/13.

45. D. L. Haviland to Sir Mark Turner, 22 March 1947. TNA: PRO BT 211/235. Regarding the claims of overclassification, the CCG/BE responded that 321/2694 final reports were classified and 2/3 not graded higher than "Restricted," a response "I think . . . supplies ammunition for a reasonably honest answer to the Russian allegations."

46. Alexander King to H. L. Verry, 26 April 1946. TNA: PRO BT 211/17.

47. Gimbel, *Science, Technology, and Reparations*, 115–34.

48. Farquharson, "Governed or Exploited?," 42.

49. R. H. Bright, Chairman, BIOS Group V, "Industrialists Visits to Germany," 30 July 1946. TNA: PRO BT 211/21.

50. R. E. Sainsbury, "Industrial Visits to Germany," 2 August 1946. TNA: PRO BT 211/21.

51. E. G. Lewin to All Regional Research Officers, "Applications by Germans for Patents in Foreign Countries," 9 August 1947. TNA: PRO FO 1032/93.

52. MacLeod, *Inventing the Industrial Revolution*, 13.

53. Hagen, "Patents Legislation and German FDI in the British Chemical Industry before 1914"; Zimmermann, *Patentwesen in Der Chemie*. Regarding American "know-how," Prime

Minister Lloyd-George objected to allowing a patent to simply be worked within the empire because it would allow Americans to work in Canada and thereby lose the benefit of America's patents in the UK. *Hansard*, HC, 9 August 1907, Vol. 180, col. 645–76.

54. "German Patents and the German Patent Office," 11 November 1947. TNA: PRO FO 1032/93.

55. "Treatment of German Trade Marks: Brief for the UK Delegation to the Council of Foreign Ministers, November 1947," 11 November 1947. TNA: PRO FO 1032/93.

56. "Treatment of German-Owned Patents," http://treaties.fco.gov.uk/docs/fullnames/pdf/1948/TS0015%20(1948)%20CMD-7359%201946%2027%20JUL,%20LONDON%3B%20TREATMENT%20OF%20GERMAN-OWNED%20PATENTS.pdf.

57. E. G. Lewin to All Regional Research Officers, "Status of the German Patent and Trade Mark Systems," 5 August 1947. TNA: PRO FO 1032/93.

58. On German patent law, see Berghahn, "West German Reconstruction and American Industrial Culture, 1945–1960"; Braun, *The German Economy in the Twentieth Century*; Gispen, *Poems in Steel*; and Patentamt, *Hundert Jahre Patentamt*.

59. [Untitled], 7 June 1947. TNA: PRO FO 1043/93; E. G. Lewin to All Regional Research Officers, "Status of the German Patent and Trade Mark Systems," TNA: PRO FO 1032/93.

60. [Untitled], 7 June 1947. TNA: PRO FO 1043/93; E. G. Lewin to All Regional Research Officers, "Status of the German Patent and Trade Mark Systems," TNA: PRO FO 1032/93.

61. [Untitled], 7 June 1947. TNA: PRO FO 1043/93; E. G. Lewin to All Regional Research Officers, "Status of the German Patent and Trade Mark Systems," TNA: PRO FO 1032/93.

62. Regional Research Office, Research Branch to Australian Political Adviser, Australian Military Mission, "Applications by Germans for Patents and Trade-Marks in Foreign Countries," 20 August 1947.

63. E. G. Lewin to All Regional Research Officers, "Status of the German Patent and Trade Mark Systems," TNA: PRO FO 1032/93.

64. "Patents-Policy," undated. TNA: PRO FO 1032/98; E. G. Lewin to President, Economic Sub-Commission, "Patent, Trade-Mark, etc. Protection for Germans in the UK," 2 October 1947. TNA: PRO FO 1032/98.

65. Control Commission for Germany (British Element), Scientific and Technical Research Board, "Re Work of Research Branch and Our Future Policy Covering Control and Encouragement of German Science," 18 June 1948. TNA: PRO BT 211/30.

66. "Patents-Policy," undated. TNA: PRO FO 1032/98.

67. Hughes, "Notes on the Prospect Confronting Post-War British Patent Property," 729.

68. Heald, "Patent Law."

69. Hughes, "Notes on the Prospect Confronting Post-War British Patent Property," 730.

70. Hughes, "Notes on the Prospect Confronting Post-War British Patent Property," 758; Spencer, "The German Patent Office."

71. Borkin and Welsh, *Germany's Master Plan*.

72. Regarding German "infiltration" into Spain, see *Hansard*, HC, 3 February 1944, Vol. 396, col. 1431W.

73. Vaughan, "Recent Patent Legislation in Great Britain." For an overview of British patent history, see especially Board, British Library, *British Patents of Invention, 1617–1977*; Gudmestad, "Patent Law of the United States and the United Kingdom"; Meshbesher, "The Role of History in Comparative Patent Law"; Pottage and Sherman, *Figures of Invention*.

74. Board, British Library, *British Patents of Invention, 1617–1977*, 7; Boehm, *The British Patent System*, 4–5.

75. Hughes, "Notes on the Prospect Confronting Post-War British Patent Property," 768.

76. *Hansard*, HC, 19 April 1944, Vol. 399, col. 216–312.

77. *Hansard*, HC, 19 April 1944, Vol. 399, col. 216–312.

78. *Hansard*, HL, 29 May 1945, Vol. 136, col. 246–59.

79. *Hansard*, HL, 29 April 1948, Vol. 155, col. 568–90. Regarding BIOS, see *Hansard*, HC, 2 March 1949, Vol. 462, col. 482–504.

80. *Hansard*, HL, 29 April 1948, Vol. 155, col. 568–90.

81. *Hansard*, HL, 29 April 1948, Vol. 155, col. 568–90.

82. *Hansard*, HL, 29 April 1948, Vol. 155, col. 568–90.

83. *Hansard*, HL, 29 April 1948, Vol. 155, col. 591–97.

84. *Hansard*, HC, 19 April 1944, Vol. 399, col. 216–312.

85. *Hansard*, HL, 30 October 1947, Vol. 152, col. 349–410.

86. *Hansard*, HC, 1 May 1956, Vol. 552, col. 215–58; *Hansard*, HC, 13 June 1956, Vol. 554, col. 659–701.

87. *Hansard*, HC, 1 May 1956, Vol. 552, col. 215–58; *Hansard*, HC, 13 June 1956, Vol. 554, col. 659–701.

88. *Hansard*, HC, 19 April 1944, Vol. 399, col. 216–312.

89. "Science Students at the Technische Hochschulen and Universities in the British Zone," January 1947. FO 1032/93.

90. BT 211/19, GD 1698/46, November 1946. Citation from Glatt, "Reparations and the Transfer of Scientific and Industrial Technology from Germany," 880.

91. Turner, *Reconstruction in Post-War Germany*, x–xi.

92. J. T. Keyd to G. S. Mansell, 28 February 1946. TNA: PRO BT 211/41.

93. *Hansard*, HL, 12 May 1953, Vol. 132, col. 395–420.

94. W. G. Glennie, Deputy Regional Controller, "Bristol Exhibition of Industrial Intelligence Work in Germany," 28 December 1946. TNA: PRO BT 211/22.

95. "Report on a Meeting of Technical Investigations in Germany by Individual Firms," July 1946. TNA: PRO BT 211/21; Derek Wood to C. S. Low, Esq., 6 March 1947. TNA: PRO BT 211/24.

Chapter 3 · French Planning for German Science

1. For nuclear science as an investment made by the French state with the conscious aim of using technology to build a new national identity, see Hecht, *The Radiance of France*. For the struggles of French science under Nazi occupation, see Chevassus-au-Louis, *Savants Sous l'Occupation*.

2. Économie Nationale, "Le Problème de la recherche technique et scientifique allemande," 10 March 1946. ANFF, CNRS records, RG 19780283, carton 35.

3. CNRS, "Le Probleme Francaise," 10 March 1946. ANFF, CNRS, RG 19780283, carton 35.

4. Économie Nationale, "Le Problème de la recherche technique et scientifique allemande," 10 March 1946. ANFF, CNRS records, RG 19780283, carton 35.

5. Regarding France's relative place in postwar diplomatic discussions, see, for example, Hitchcock, *France Restored*, and Cairncross, *The Price of War*.

6. F. A. A. Blake, Col., General Staff (GS), G-2, to Office of the Assistant Chief of Staff, G-2, SHAEF, 25 June 1945. RG 260.3.6, Vol. 17, Box 2, folder 8. NACP.

7. F. A. A. Blake, Col., GS, G-2, to Lt. Col. Stagnaro, le Chef de la 2eme Section, d'État-major General de la Défense Nationale, "German Technical Intelligence in France," 31 January 1945. RG 260.3.6, Vol. 17, Box 2, folder 8. NACP.

8. A. S. Knight, Col., GSC, Chief "T" Sub-Division, G-2, to G-2, SHAEF Mission (France), "German Technical Intelligence in France," 24 January 1945. RG 260.3.6, Vol. 17, Box 2, folder 8. NACP.

9. J. J. Davis, Brigadier General, Adjutant General, to Head, SHAEF Mission (France), "French Participation in the Collection of Technical Information in Germany," 15 June 1945. RG 260.3.6, Vol. 17, Box 2, folder 8. NACP.

10. A. Juin, Gen., Chief, EMGDN, to General Commanding, SHAEF Mission (France),

"General Technical Intelligence in France," 28 February 1945. RG 260.3.6, Vol. 12, Box 2, folder 8. NACP.

11. Hayes, *Industry and Ideology*, 278–90.

12. Combined Chiefs of Staff, Washington, DC, to Chief of the French Military Mission. RG 260.3.6, Vol. 12, Box 2, folder 8. NACP.

13. J. J. Davis, Brig. Gen., AG, to Head, SHAEF Mission (France), "French Participation in the Collection of Technical Information in Germany," 15 June 1945. RG 260.3.6, Vol. 17, Box 2, folder 8. NACP.

14. "Report on French Participation in the Collection of Technical Intelligence in Germany," 24 June 1945. RG 260.3.6, Vol. 17, Box 2, folder 8. NACP.

15. P. M. Wilson, Major, GS, Enemy Personnel Exploitation Section, G-2, FIAT to Intelligence Section, Exploitation Division, United States Strategic and Tactical Air Forces, "Exploitation of German Personalities in French Zone," 8 June 1945. RG 260.3.6, Vol. 17, Box 2, folder 8. NACP.

16. H. Read, Lt. Col., to R. J. Maunsell, Brig. Gen., Chief FIAT (Britain), "Exchange of Information with the French," 20 August 1945. RG 260.3.6, Vol. 17, Box 1, folder 30. NACP.

17. Deputy Military Governor (US) for Germany to Joint Chiefs of Staff, 21 November 1945. RG 260.3.6, Vol. 13, Box 3, folder 11. NACP.

18. A. Juin, Gen., Chief, EMGDN, to General Commanding, SHAEF Mission (France), "General Technical Intelligence in France," 28 February 1945. RG 260.3.6, Vol. 12, Box 2, folder 8. NACP.

19. A. S. Knight, Col., GSC, Chief "T" Sub-Division, G-2, to G-2, SHAEF Mission (France), "German Technical Intelligence in France," 24 January 1945, and "Report on French Participation in the Collection of Technical Intelligence in Germany," 24 June 1945. Both in RG 260.3.6, Vol. 17, Box 2, folder 8. NACP.

20. In the original French: "Il y aura lieu de faire transférer en France les scientifiques ou techniciens allemands de grande valeur pour les faire interroger à loisir sur leurs travaux et éventuellement les engager à rester à notre disposition." SHAT/EMGDN, 5 P 118. Reference originally found in Ludmann-Obier, "Un Aspect de la Chasse aux Cerveaux."

21. Eugene L. Harrison, Brig. Gen., GSC, G-2, to "T" Sub-Division, G-2, SHAEF Mission (France), "Activities of Sécurité militaire," 2 June 1945. RG 260.3.6, Vol. 17, Box 2, folder 8. NACP.

22. "Report on French Participation in the Collection of Technical Intelligence in Germany," 24 June 1945. RG 260.3.6, Vol. 17, Box 2, folder 8. NACP.

23. "Report on French Participation in the Collection of Technical Intelligence in Germany," 24 June 1945. RG 260.3.6, Vol. 17, Box 2, folder 8. NACP; FIAT (US) to Enemy Personnel Exploitation Section, FIAT (Britain), 8 February 1946. RG 260.3.6, Vol. 17, Box 1, folder 30. NACP; A. E. O'Flaherty Jr., Col., Infantry, Chief, Operations Branch, FIAT (US), to Office of the Assistant Chief of Staff, G-2, Operations Branch, Technical Intelligence Unit, US Forces European Theater, "Evacuation of German Scientists to French Zone," 28 May 1946. RG 260.3.6, Vol. 17, Box 1, folder 30. NACP.

24. Gimbel, *Science, Technology, and Reparations*, 41–42.

25. Gimbel, *Science, Technology, and Reparations*, 46–47.

26. Gimbel, *Science, Technology, and Reparations*, 47.

27. Ludmann-Obier, "Un Aspect de la Chasse aux Cerveaux," 196–97.

28. Ludmann-Obier, "Un Aspect de la Chasse aux Cerveaux," 198.

29. Villain, "France and the Peenemunde Legacy," 124.

30. Villain, "France and the Peenemunde Legacy," 127.

31. Kolodziej, *Making and Marketing Arms*, 43.

32. F. Canac, Directeur du Centre de Recherches S.I.M. to Le Directeur-Adjoint du Centre

National de la Recherche Scientifique, "Objet: Laboratoire Forster," 18 April 1946. ANFF, CNRS records, RG 19780283, carton 35.

33. The Import/Export Service had a monopoly on importations to France, the Commercial Mission had a monopoly on commercial contracts between France and Germany, the Office for External Commerce had a monopoly on commercial operations in the French occupation zone, and possibly more had similarly overlapping "sole" authority. Caigniard, Lt. Col., Section d'information scientifique (FIAT [France]), Chef de la Sous-Section CNRS, "Note pour le Centre National de la Recherche Scientifique," 4 July 1946. ANFF, CNRS records, RG 19780283, carton 35.

34. "Services Francais avec lesquels la Mission CNRS en Allemagne est en Rapport," July 1949. ANFF, CNRS records, RG 19780284, carton 115.

35. For a fuller history of the organizational ties of the CNRS mission to Germany, see Defrance, "La mission du CNRS en Allemagne (1945–1950)"; Ludmann-Obier, "La Mission du CNRS en Allemagne (1945–1950)."

36. M. Decombre to High Commissioner, French Occupation Government in Germany, "Rattachement de la mission CNRS en Allemagne," 7 October 1949. ANFF, CNRS records, RG 19780284, carton 115. See also Defrance, "La mission du CNRS en Allemagne (1945–1950)"; Guthleben, "La participation du Centre à l'effort scientifique de guerre"; Ludmann-Obier, "La Mission du CNRS en Allemagne (1945–1950)"; and Prost, "Les Origines des Politiques de la Recherche en France (1939–1958)."

37. Économie Nationale, "Le Problème de la recherche technique et scientifique allemande," 10 March 1946. ANFF, CNRS records, RG 19780283, carton 35.

38. "Note complementaire sur les conditions d'un controle de la Recherche Scientifique en Allemagne dans le cadre du traité de paix," 7 March 1946. ANFF, CNRS records, RG 19780283, carton 35.

39. In French, "un savant ou un technicien déplace dans un autre pays . . . est pratiquement rendu stérile." "Note complementaire sur les conditions d'un controle de la Recherche Scientifique en Allemagne dans le cadre du traité de paix," 7 March 1946. ANFF, CNRS records, RG 19780283, carton 35.

40. "L'intérêt du transfert est pratiquement nul lorsqu'il existe en France un organisme analogue à celui qui existe en Allemagne. . . . Un transfert, en coupant tous ces liens, dévaloriserait fortement le centre allemand. Il . . . no reprendrait de dynamisme qu'après avoir rétabli de nouveaux liens avec les milieux scientifiques français. . . . Il est alors préférable de laisser subsister le centre allemand et de contrôler son activité de manière à retirer de l'état d'avancement de la recherche scientifique et technique en Allemagne." "Note complementaire sur les conditions d'un controle de la Recherche Scientifique en Allemagne dans le cadre du traité de paix," 7 March 1946. ANFF, CNRS records, RG 19780283, carton 35.

41. "En règle générale, le transfert des laboratoires et centres de recherches allemands amènera une diminution considérable du rendement de ces laboratoires qui seront d'une intégration difficile et peu rentable dans l'économie française." "Note complementaire sur les conditions d'un controle de la Recherche Scientifique en Allemagne dans le cadre du traité de paix," 7 March 1946. ANFF, CNRS records, RG 19780283, carton 35.

42. "Les principes énoncés sont excellents: en particulier le problème national est parfaitement pose." "Procès-Verbal de la 3eme Séance du Comité de coordination scientifique de Défense Nationale, Réunion Spéciale, a la Grande Chancellerie de la Légion d'Honneur," 25 March 1946. ANFF, CNRS records, RG 19780283, carton 35.

43. In French: Il "estime que le transport des établissements de recherche en France diminuera considérablement leur rendement. Les chercheurs travaillent en _équipes_ et il est vain d'espérer que quelques techniciens prélevés sur un établissement pourront continuer leur travaux dans de bonnes conditions" (emphasis original). "Procès-Verbal de la 3eme Séance du Comité de

coordination scientifique de Défense Nationale, Réunion Spéciale, a la Grande Chancellerie de la Légion d'Honneur," 25 March 1946. ANFF, CNRS records, RG 19780283, carton 35.

44. CNRS, "Note sue le controle de la recherche allemande," date unknown. ANFF, CNRS records, RG 19780283, carton 35.

45. CNRS, "Note sue le controle de la recherche allemande," date unknown. ANFF, CNRS records, RG 19780283, carton 35.

46. Économie Nationale, "Le Problème de la recherche technique et scientifique allemande," 10 March 1946. ANFF, CNRS records, RG 19780283, carton 35.

47. "Controle des Rescherches Scientifiques et Techniques tant dans les établissements d'enseignement que dans laboratoires particuliers et les Establissements Industriels," 7 March 1946. ANFF, CNRS records, RG 19780283, carton 35.

48. Économie Nationale, "Le Problème de la recherche technique et scientifique allemande," 10 March 1946. ANFF, CNRS records, RG 19780283, carton 35.

49. CNRS, "Le Probleme Francaise," 10 March 1946. ANFF, CNRS, RG 19780283, carton 35.

50. "Procès-Verbal de la réunion de la sous-Commission des Relations avec l'Etranger," 14 December 1946. ANFF, CNRS records, RG 19780283, carton 24.

51. "Ecole Superiore des Industries Chemiques to Joliot et Tessier, Direction du CNRS," 19 April 1945. ANFF, CNRS record, RG 19780283, carton 35.

52. "Stagiaires, qui a pour but officiel de se former aux méthodes allemandes, mais pour but officieux de 'pomper', tout ce qu'ils peuvent dans les laboratoires allemandes … ils n'ont pris aucun engagement de ce genre." "Comte rendu de la 1ere reunion de la commission superieure de la recherche scientifique en allemagne, tenue le 3.3.47 a Baden," 3 March 1947. ANFF, CNRS records, RG 19780283, carton 36.

53. "Reunion de la commission superieure de la recherche scientifique en allemagne en date du 13 Octobre 1947," 13 October 1947. ANFF, CNRS records, RG 19780283, carton 36.

54. Eck, "Les Contacts Entre Groupes de l'industrie Chimique Français et Allemands de 1945 à la Fin des Années 1960: Entre Compétition et Coopération."

55. Bossaut, "Armaments et Relations Franco-Allemandes (1945–1963): Coopération et Recherche de Puissance," 156.

56. Commission des Relations avec l'Etranger, "Proces-Verbal de la réunion du samedi 29 novembre 1947," 29 November 1947. ANFF, CNRS records, RG 09780283, carton 24.

57. Krige, *American Hegemony and the Postwar Reconstruction of Science in Europe.*

58. On this, see, for example, Busquin, "Le Rôle des Réseaux de Scientifiques Dans l'émergence d'un Espace Européen de la Recherche"; Hitchcock, *France Restored*; Kohlrausch and Trischler, *Building Europe on Expertise*; and Krige, "The Peaceful Atom as Political Weapon."

59. "Comte Rendu sue la mission du CNRS en Allemagne," 25 November 1946. ANFF, CNRS records, RG 19780284, carton 115.

60. Decombe, Chef de la Sous-Section CNRS, FIAT (France) to M. Autheman (Decombe to Autheman), "A l'attention de Monsieur Autheman," 2 February 1950. ANFF, CNRS records, RG 19780284, carton 115.

61. Using conversions on http://www.historicalstatistics.org/Currencyconverter.html, an imperfect data source but the best available.

62. Gimbel, *Science, Technology, and Reparations.* See especially 94–114.

63. As evidence of this lack of funding, Jean Lecomte, director of research for CNRS, was sent on a fact-finding mission to America and Canada in 1947 to learn of advances in infrared technology. Though he claims significant advances came from this trip, each of his several reports complains bitterly of not having enough money to visit with scientists in off hours, eat well, or travel comfortably, damaging the networking benefits of the trip. Jean Lecomte, Directeur de Recherches au CNRS, "Rapport presente a la Direction generale des Relations culturelles," November 1947. ANFF, CNRS records, RG 19780284, carton 56. Of course, his claims

of great value obtained from these trips should be taken at face value no more than the claims of enormous value from "intellectual reparations" examined elsewhere in this study.

64. ANFF, CNRS records, RG 19780284, cartons 57–60; Pestre, "Guerre, Renseignement Scientifique et Reconstruction, France, Allemagne et Grande-Bretagne dans les Années 1940."

65. Derek Wood, to A. Percival, Esq., British Embassy. 29 May 1947. BNA, PRO, BT 211, carton 24.

66. "Report on Visit to Hoechst, Offenburg, Baden-Baden and Paris, 21–27 July 1946. BIOS Trip No. 2575," 27 July 1946. BNA, PRO, BT 211, carton 17.

67. Judt, "Exploitation by Integration?," in Judt and Ciesla, *Technology Transfer out of Germany after 1945*, 32.

68. Trevor Evans, "Smith and the Secrets of Schmidt," *Daily Express*, 9 October 1945.

69. "Re: Statistics of Clearances of French and British Technical Investigators for Period 0830 Hours July 29 to 1200 Hours," 24 August 1946. RG 260.3.6, Vol. 17, Box 1, folder 15. NACP.

70. Board of Trade, "BIOS Exhibition: Notes of Telephone Conversation with Professor Linsted (Formerly Chairman of CIOS and, Later of BIOS)," 27 November 1946. BNA, PRO, BT 211, carton 22.

71. American and British policy demands explanation in itself, of course, rather than being some kind of "natural" course from which the French truly "diverged." The intellectual origins of the Anglo-American policies are beyond the scope of this chapter, and to the extent that the topic here is the differences between their approach and the French one, either phrasing communicates the concept. Board of Trade, "BIOS Exhibition: Notes of Telephone Conversation with Professor Linsted (Formerly Chairman of CIOS and, Later of BIOS)," 27 November 1946. BNA, PRO, BT 211, carton 22; Board of Trade, "Employment of German Scientists and Technicians in Civil Industry in the United Kingdom," 28 August 1945. British National Archives, PRO, BT 211, carton 46.

72. Blum-Picard, "Note pour Messieurs les Directeurs," 22 October 1945. Archives National Paris, Entry F12/10022.

73. See, for example, Milward, *The Reconstruction of Western Europe, 1945–51*; Gillingham, *Coal, Steel, and the Rebirth of Europe, 1945–1955*; and Buchheim, *Die Wiedereingliederung Westdeutschlands in Die Weltwirtschaft 1945–1958*.

Chapter 4 · *Soviet Reparations and the Seizure of German Science and Technology*

1. On Soviet science, see Holloway, *Stalin and the Bomb*; Kojevnikov, *Stalin's Great Science*; Krementsov, *Stalinist Science*; Graham, *Science in Russia and the Soviet Union*; and Pollock, *Stalin and the Soviet Science Wars*.

2. A nonexhaustive, alphabetical list of main sources includes Albrecht, Heinemann-Grüder, and Wellmann, *Die Spezialisten*; Holloway, *Stalin and the Bomb*; Karlsch, *Allein Bezahlt?*; Karlsch, Laufer, and Sattler, *Sowjetische Demontagen in Deutschland 1944–1949*; Mick, *Forschen für Stalin*; Naimark, *The Russians in Germany*; Siddiqi, "Germans in Russia"; Slaveski, *The Soviet Occupation of Germany*; and Sokolov, *Soviet Use of German Science and Technology, 1945–1946*.

3. In comparison, the states of Texas and Florida together contain less than 13 percent of the United States' population in 2016. Holloway, *Stalin and the Bomb*, 144.

4. As a starting point, see the essays in Heer and Naumann, eds., *War of Extermination*.

5. Churchill to Roosevelt, 24 November 1944 (quoted in PRO CAB 127/272). Cited in Cairncross, *The Price of War*, 13.

6. Slaveski, *The Soviet Occupation of Germany*; Naimark, *The Russians in Germany*, 69–140.

7. Cairncross, *The Price of War*, 195–202.

8. Cairncross, *The Price of War*, 195–96.

9. Naimark, *The Russians in Germany*, 25.

10. Fisch, "Reparations and Intellectual Property"; Karlsch, Laufer, and Sattler, *Sowjetische Demontagen in Deutschland 1944–1949*.

11. Naimark, *The Russians in Germany*, 26–28.

12. Cairncross, *The Price of War*, 199.

13. Slaveski, *The Soviet Occupation of Germany*, 14–15.

14. Naimark, *The Russians in Germany*, 169.

15. Stokes, *Constructing Socialism*, 20; Stokes, "Assessing the Damages."

16. Naimark, *The Russians in Germany*, 26.

17. Cairncross, *The Price of War*, 200.

18. Cairncross, *The Price of War*, 200.

19. Slusser, ed., *Soviet Economic Policy in Postwar Germany*, 14–17.

20. Naimark, *The Russians in Germany*, 179–82.

21. Slusser, ed., *Soviet Economic Policy in Postwar Germany*, 41.

22. Naimark, *The Russians in Germany*, 180–81.

23. Slusser, ed., *Soviet Economic Policy in Postwar Germany*, 41.

24. Sutton, *Western Technology and Soviet Economic Development, 1945 to 1965*, 325.

25. Uhl, *Stalins V-2*; Albrecht, Heinemann-Grüder, and Wellmann, *Die Spezialisten*; Albrecht and Nikutta, *Die Sowjetische Rüstungsindustrie*.

26. Naimark, *The Russians in Germany*, 210–11.

27. Naimark, *The Russians in Germany*, 215.

28. Naimark, *The Russians in Germany*, 215–18.

29. Naimark, *The Russians in Germany*, 219.

30. Siddiqi, "Germans in Russia," 123.

31. Naimark, *The Russians in Germany*, 218.

32. Walker, *Nazi Science*.

33. Naimark, *The Russians in Germany*, 209. Citing USPOLAD, "Transfer of Scientific and Military Research Institutions from Germany to Russia," July 17, 1945, 1–2, United States National Archives, Record Group 59, 740.00119, Control (Germany), 7–1745.

34. Gimbel, "German Scientists, United States Denazification Policy, and the 'Paperclip Conspiracy'"; Jacobsen, *Operation Paperclip*.

35. Examples of these memoirs include Riehl, *Zehn Jahre im Goldenen Käfig*.

36. From Siddiqi, "Germans in Russia," 120n1.

37. Siddiqi, "Germans in Russia." Siddiqi's other work is also well worth reading, including Siddiqi, *The Red Rockets' Glare*; Siddiqi, *Sputnik and the Soviet Space Challenge, Vol. 1*; Siddiqi, *Sputnik and the Soviet Space Challenge, Vol. 2*.

38. Holloway, *Stalin and the Bomb*, 221.

39. Slaveski, *The Soviet Occupation of Germany*.

40. Cairncross, *The Price of War*, 211.

41. Mick, *Forschen für Stalin*, 15–17. Another source for estimates of how many people were involved (fewer than 3,500) can be found in Albrecht, Heinemann-Grüder, and Wellmann, *Die Spezialisten*.

42. For details of Osoaviakhim, see especially Naimark, *The Russians in Germany*, 220–28.

43. Naimark, *The Russians in Germany*, 221.

44. Maddrell, *Spying on Science*, 31. Maddrell cites Minute by P. Dean, 2/8/1946, FO 371/55906.

45. Foust, "Probe Red Kidnapping Raids: Scientist Escapes, Tells of Seizure," *Chicago Daily Tribune*, 25 October 1946.

46. Quoted in Lasby, *Project Paperclip*, 183.

47. Heisenberg did advise some to go or stay in the Eastern Zone because of overcrowding

in the Western physics job market. This story comes from Walker, *German National Socialism and the Quest for Nuclear Power, 1939–1949*, 185.

48. Cairncross, *The Price of War*, 200.

49. Naimark, *The Russians in Germany*, 187, 194.

50. Kuenzel, "Verwaltung Sowjetische [Staatliche] Aktiengesellschaften in Deutschland (SAG)," in Foitzik et al., *SMAD-Handbuch*.

51. Naimark, *The Russians in Germany*, 191.

52. Naimark, *The Russians in Germany*, 190; Cairncross, *The Price of War*, 205.

53. Radkau, "Revoltierten die Produktivkräfte Gegen den Real Existierenden Sozialismus?"

54. Stokes, *Constructing Socialism*, 29.

55. Naimark, *The Russians in Germany*, 230–33.

56. Naimark, *The Russians in Germany*, 231.

57. Naimark, *The Russians in Germany*, 231–32.

58. Naimark, *The Russians in Germany*, 232.

59. Naimark, *The Russians in Germany*, 232.

60. At least 10–11 percent can be verified to have fled West according to East German records; Steiner suggests the 25 percent figure as a good approximation for the total in Steiner, "The Return of German 'Specialists' from the Soviet Union to the German Democratic Republic," 126.

61. Steiner, "The Return of German 'Specialists' from the Soviet Union to the German Democratic Republic," 119–20.

62. The details in this paragraph are from Steiner, "The Return of German 'Specialists' from the Soviet Union to the German Democratic Republic."

63. On this topic, see Augustine, *Red Prometheus*; Geyer, "Industriepolitik in her DDR."

64. Steiner, "The Return of German 'Specialists' from the Soviet Union to the German Democratic Republic," 127.

65. Bericht über die zurückkehrenden SU-Specialisten, 31 December 1954, in SAPMO, SED J IV 2/202/56. Citation from Steiner, "The Return of German 'Specialists' from the Soviet Union to the German Democratic Republic," 124.

66. Maddrell, *Spying on Science*.

67. In some cases, this even extended to assassinating scientists. In 1943, the British Royal Air Force bombed the living quarters of the Peenemünde aeronautics facility developing the V-2 rocket, killing 130; in 1944, the US Office of Secret Services (predecessor to the CIA) sent an agent to a lecture by Werner Heisenberg with orders to assassinate him if he indicated progress on atomic bombs. See Jones, *Most Secret War*, 346, and Richelson, "When Kindness Fails," 248–49.

68. Steiner, "The Return of German 'Specialists' from the Soviet Union to the German Democratic Republic," 126.

69. Quoted in Steiner, "The Return of German 'Specialists' from the Soviet Union to the German Democratic Republic," 120.

70. Augustine, "Wunderwaffen of a Different Kind," 582; Steiner, "The Return of German 'Specialists' from the Soviet Union to the German Democratic Republic," 126.

71. The phrase comes from Hans Becker, "a former employee of the microelectronics enterprise Arbeitsstelle für Molekularelektronic," in interview with Dolores Augustine. See Augustine, "Wunderwaffen of a Different Kind," 580.

72. Augustine, "Wunderwaffen of a Different Kind," 585.

73. Nettl, "German Reparations in the Soviet Empire."

74. Quote in Lasby, *Project Paperclip*, 6.

75. Cited in Naimark, *The Russians in Germany*, 232, as "Otchet o rabote otd. nauki i tekhniki," GARF, f. 7184, op. 1, d. 148, l. 13.

76. Though the Soviet Union heavily pressured its vassal states to take on its system, there was some important variety within the Soviet bloc. Czechoslovakia, for example, retained a patent system developed before the 1940s until the 1970s. See Trimble, "The Patent System in Pre-1989 Czechoslovakia."

77. Stokes, *Constructing Socialism*, 34. See also Karlsch, *Allein Bezahlt?*

78. Naimark, *The Russians in Germany*; Stokes, *Constructing Socialism*; Karlsch, *Allein Bezahlt?*

Chapter 5 · *Academic Science and the Reconstruction of Germany*

1. Of course, there is no real, clear distinction between "academic" and "industrial" or "basic" and "applied" science. Much industrial research takes places on university campuses by university professors, and often research undertaken by corporate researchers in industrial laboratories is important for university-based researchers. Still, though the categories overlap and I will not try to force and clear lines between them, this chapter looks more at research not intended for immediate economic pay-off or application.

2. On the different perceptions of science and governance in this era, see, for example, Beyler, "Hostile Environmental Intellectuals?"; Brown, *Science in Democracy*; Jasanoff, *Designs on Nature*; Kevles, "The National Science Foundation and the Debate over Postwar Research Policy, 1942–1945"; Kevles, *The Physicists*; and Layton, "Mirror-Image Twins."

3. Published in *Official Gazette of the Control Council for Germany*, 6 (30 April 1946): 132–43. Citation from Cassidy, "Controlling German Science, I," 200.

4. Cassidy, "Controlling German Science, I," 225.

5. Abelshauser, *The Dynamics of German Industry*; Abelshauser, *German Industry and Global Enterprise*.

6. On the development of these institutions, see Kohler, *Partners in Science*, and Solovey, *Shaky Foundations*.

7. Macrakis, *Surviving the Swastika*, 110–11.

8. Macrakis, *Surviving the Swastika*.

9. Oberkrome, Orth, and Tagung, *Die Deutsche Forschungsgemeinschaft 1920–1970*.

10. See, for example, Daston and Galison, *Objectivity*, and Bowler, *Science for All*.

11. Kramer, "British Dismantling Politics, 1945–1949," 126.

12. On British occupation policy more generally, see Deighton, *The Impossible Peace*; Foschepoth and Steininger, *Die Britische Deutschland- und Besatzungspolitik 1945–1949*; Schneider, "Britische Besatzungspolitik 1945"; Edgerton, *Warfare State*; Meehan, *A Strange Enemy People*; Glatt, "Reparations and the Transfer of Scientific and Industrial Technology from Germany."

13. Control Commission for Germany (British Element), Scientific and Technical Research Board, "Re Work of Research Branch and Our Future Policy Covering Control and Encouragement of German Science," 18 June 1948. TNA: PRO BT 211/30.

14. Control Commission for Germany (British Element), Scientific and Technical Research Board, "Re Work of Research Branch and Our Future Policy Covering Control and Encouragement of German Science," 18 June 1948. TNA: PRO BT 211/30.

15. Control Commission for Germany (British Element), Scientific and Technical Research Board, "Re Work of Research Branch and Our Future Policy Covering Control and Encouragement of German Science," 18 June 1948. TNA: PRO BT 211/30.

16. Control Commission for Germany (British Element), Scientific and Technical Research Board, "Re Work of Research Branch and Our Future Policy Covering Control and Encouragement of German Science," 18 June 1948. TNA: PRO BT 211/30.

17. On the refounding of the Max Planck Gesellschaft, see Carson, *Heisenberg in the Atomic Age*; Heilbron, *The Dilemmas of an Upright Man*; Macrakis, *Surviving the Swastika*; Sachse, "'Whitewash Culture'"; Vierhaus and vom Brocke, eds., *Forschung im Spannungsfeld von Politik*

und Gesellschaft; vom Brocke and Laitko, eds., *Die Kaiser-Wilhelm-/Max-Planck-Gesellschaft und ihre Institute.*

18. Macrakis, *Surviving the Swastika*, 190.

19. Macrakis, *Surviving the Swastika*, 191.

20. Macrakis, *Surviving the Swastika*, 193.

21. Ministere de l'education nationale, "Directives Generales pour l'organisation er le contrôle de la recherche allemande," 1 June 1946. ANFF, CNRS records, RG 19780283, carton 35.

22. Scientific and Technical Research Board, "Review of Scientific & Technical Research in Germany since May, 1945," 15 January 1947. TNA: PRO BT 211/30.

23. Scientific and Technical Research Board, "Minutes of 11th Meeting," 13 January 1947. TNA: PRO BT 211/30.

24. Scientific and Technical Research Board, "Review of Scientific & Technical Research in Germany since May, 1945," 15 January 1947. TNA: PRO BT 211/30.

25. Control Commission for Germany (British Element), Scientific and Technical Research Board, "Re Work of Research Branch and Our Future Policy Covering Control and Encouragement of German Science," 18 June 1948. TNA: PRO BT 211/30.

26. The relationship between science and empire is a long and important one in British science. See Hecht, *Entangled Geographies*; Hodge and Bennett, eds., *Science and Empire*; and Jasanoff, "Biotechnology and Empire."

27. Control Commission for Germany (British Element), Scientific and Technical Research Board, "Re Work of Research Branch and Our Future Policy Covering Control and Encouragement of German Science," 18 June 1948. TNA: PRO BT 211/30.

28. For a starting place on the role of scientists in American society in this era, see Jewett, *Science, Democracy, and the American University*; Kevles, "Cold War and Hot Physics"; Kevles, *The Physicists*; Leslie, *The Cold War and American Science.*

29. Hershberg, *James B. Conant*; Kaiser, "Cold War Requisitions, Scientific Manpower, and the Production of American Physicists after World War II"; Kaiser, "Nuclear Democracy"; Kevles, "The National Science Foundation and the Debate over Postwar Research Policy, 1942–1945"; Wang, *In Sputnik's Shadow.*

30. For more on the Morgenthau Plan and the broader history of occupation policy, see Balabkins, *Germany under Direct Controls*; Cairncross, *The Price of War*; and Gimbel, *The Origins of the Marshall Plan.*

31. Examples include Backer, *Priming the German Economy*; Gaddis, *The United States and the Origins of the Cold War*; Milward, *The Reconstruction of Western Europe, 1945–51*; and Junker, ed., *The United States and Germany in the Era of the Cold War, 1945–1968.*

32. This summary derives principally from Cassidy, "Controlling German Science, I," 220–22. For more on Adams, see also Doel, "Roger Adams," and Tarbell, *Roger Adams.*

33. Cassidy, "Controlling German Science, I."

34. "Meeting Held in FIAT Conference Room at HOCHST at 1030 Hours on Wednesday 15th May 1946 to Consider Bearing on the Commission of War Crimes by German Scientists Believed to be Guilty of Inhuman Experimentation on Living Men and Women," 15 May 1946. RG 260, box 17/01, folder 26, NACP.

35. "P&I/ECON/2220: Technical and Scientific Research in Germany after the War," 19 June 1945. RG 260, box 17/04, folder 26, NACP.

36. Krige, *American Hegemony and the Postwar Reconstruction of Science in Europe.*

37. Krige, *American Hegemony and the Postwar Reconstruction of Science in Europe*, 39.

38. Krige, *American Hegemony and the Postwar Reconstruction of Science in Europe*, 10.

39. Examples in the historiography include Cairncross, *The Price of War*; Craig and Loewenheim, *The Diplomats, 1939–1979*; and Hitchcock, *France Restored.*

40. On the history of the CNRS, see Bourquin, "Le Comité National de la Recherche Scien-

tifique"; Charpentier-Morize, "La Contribution des 'Laboratoires Propres' du CNRS à la Recherche Chimique en France de 1939 à 1973"; Dosso, "Louis Rapkine (1904–1948) et la Mobilisation Scientifique de la France Libre"; Ludmann-Obier, "La Mission du CNRS en Allemagne (1945–1950)"; Pestre, "Guerre, Renseignement Scientifique et Reconstruction, France, Allemagne et Grande-Bretagne dans les Annees 1940," in Guillerme, *De la Diffusion des Sciences à l'espionnage Industriel, XVe–XXe Siècle*; Picard, "La Longue Marche vers le CNRS (1901–1945)"; Prost, "Les Origines des Politiques de la Recherche en France (1939–1958)"; and Zallen, "The Rockefeller Foundation and French Research."

41. This summary draws primarily from Day, "Science, Applied Science and Higher Education in France 1870–1945, an Historiographical Survey since the 1950s." More recent work on the development of French science and its relationship to the state includes Boncourt, "L'internationalisation de la science politique"; Griset and Bouvier, "De l'histoire des techniques à l'histoire de l'innovation"; Tournès, "La Fondation Rockefeller et la Construction d'une Politique des Sciences Sociales en France (1918–1940)." There are many additional articles of interest in *La revue pour l'Histoire du CNRS*.

42. Day, "Science, Applied Science and Higher Education in France 1870–1945, an Historiographical Survey since the 1950s," 376.

43. Picard, "La Longue Marche vers le CNRS (1901–1945)," 7.

44. Prost, "Les Origines des Politiques de la Recherche en France (1939–1958)."

45. Prost, "Les Origines des Politiques de la Recherche en France (1939–1958)," 3.

46. P. Donzelot to F. Joliot and G. Tessier, 19 April 1945. ANFF, CNRS records, RG 19780283, carton 35.

47. Loosely translated, "the heritage of the laboratory, material or spiritual."

48. In French: "nous soyions à même de prendre sur [Allemagne] la supériorité à laquelle nous prétendons." Économie nationale, "Le Problème de la recherche technique et scientifique allemande," 10 March 1946. ANFF, CNRS records, RG 19780283, carton 35.

49. "Reorganisation du Centre National de la Recherche Scientifique," 7 Sept 1944. ANFF, CNRS records, RG 19780284, carton 55.

50. "Organisation de la Recherche en Allemagne," 4 April 1945. ANFF, CNRS records, RG 19780283, carton 35.

51. "Note par le Comité de coordination scientifique de defense nationale," 9 November 1945. ANFF, CNRS records, RG 19780284, carton 56.

52. Économie Nationale, "Le Problème de la recherche technique et scientifique allemande," 10 March 1946. ANFF, CNRS records, RG 19780283, carton 35.

53. Économie Nationale, "Le Problème de la recherche technique et scientifique allemande," 10 March 1946. ANFF, CNRS records, RG 19780283, carton 35.

54. "Note sur l'organisation de la recherche scientifique et technique en allemagne," undated. ANFF, CNRS records, RG 19780283, carton 35.

55. "Note sur l'organisation de la recherche scientifique et technique en allemagne," undated. ANFF, CNRS records, RG 19780283, carton 35.

56. "Note sur l'organisation de la recherche scientifique et technique en allemagne," undated. ANFF, CNRS records, RG 19780283, carton 35.

57. "PI/ECON/2037: Technical and Scientific Research in Germany after the War," undated 1945. RG 260, box 17/03, folder 16, NACP.

58. "Note sur l'organisation de la recherche scientifique et technique en allemagne," undated. ANFF, CNRS records, RG 19780283, carton 35.

59. "Directives Generales pour l'organisation et le contrôle de la recherche allemande," 1 June 1946. ANFF, CNRS records, RG 19780283, carton 35.

60. Roughly, "a policy of dispersion." J. Peres, "NOTE pour le Colonel CAGNIARD," undated. ANFF, CNRS records, RG 19780283, carton 35.

61. Centre National de la Recherche Scientifique, "Compte Rendu sur la mission du C.N.R.S. en Allemagne," 25 November 1946. ANFF, CNRS records, RG 19780284, carton 115.

62. Le Sous-Lieutenant Riviere, "Note sue la mission scientifique—C.N.R.S. Vienne," 22 July 1946. ANFF, CNRS records, RG 19780283, carton 35.

63. Gordin, *Scientific Babel.*

64. "Note sue l'organisation de la recherche scientifique et technique en allemagne," 10 March 1946. ANFF, CNRS records, RG 19780283, carton 35.

65. Le Conseiller Culturel à Monsieur le Directeur de l'Enseigne Technique, 11 January 1947. ANFF, CNRS records, RG 19780283, carton 24; Comité Superieur des Etudes Techniques de Chimie en Allemange, "Procès-verbal de la reunion tenue le 17 Octobre 1947, et laboratoire central des services chimiques de l'état," 17 October 1947. ANFF, CNRS records, RG 19780283, carton 36.

66. Comité Superieur des Etudes Techniques de Chimie en Allemange, "Procès-verbal de la reunion tenue le 17 Octobre 1947, et laboratoire central des services chimiques de l'état," 17 October 1947. ANFF, CNRS records, RG 19780283, carton 36.

67. "Commission des relations avec l'étranger, procès-verbal de la reunion du samedi 29 novembre 1947 a 9 h. 30 au C.N.R.S.," 29 November 1947. ANFF, CNRS records, RG 19780283, carton 24.

68. "Procès-verbal de la réunion de la sous-Commission des Relations avec l'Étranger," 14 December 1946. ANFF, CNRS records, RG 19780283, carton 24.

69. "Commission des relations avec l'étranger, procès-verbal de la reunion du samedi 29 novembre 1947 a 9 h. 30 au C.N.R.S.," 29 November 1947. ANFF, CNRS records, RG 19780283, carton 24.

70. CNRS, "Note sure le contrôle de la recherche allemande," undated. ANFF, CNRS records, RG 19780283, carton 35.

71. "Memorial: Concerning the Present Situation of the German Scientists and Technical Experts Evacuated from the Thuringia Area by the US Army," January 1946. RG 260, box 17/11, folder 1, NACP.

72. Scientific and Technical Research Board, "Review of Scientific & Technical Research in Germany since May, 1945," 15 January 1947. TNA: PRO BT 211/30.

73. Scientific and Technical Research Board, "Review of Scientific & Technical Research in Germany since May, 1945," 15 January 1947. TNA: PRO BT 211/30.

74. Walker, *German National Socialism and the Quest for Nuclear Power, 1939–49*, 210–12.

75. "FIAT Review of Nuclear Physics," 24 June 1947. RG 260, box 17/02, folder 10, NACP.

76. Ralph Osborne to Gen. de Verbigier de St. Paul, 27 January 1947. RG 260, box 17/05, folder 19, NACP.

77. Ralph Osborne to Gen. de Verbigier de St. Paul, 27 January 1947. RG 260, box 17/05, folder 19, NACP.

78. Carson, *Heisenberg in the Atomic Age*; Sachse, "'Whitewash Culture'"; Walker, *Nazi Science.*

79. Krige, *American Hegemony and the Postwar Reconstruction of Science in Europe.*

Chapter 6 · *Documentation and Information Technology*

1. This and the following sentence draw from Cozzens, "The Discovery of Growth."

2. Price, *Science since Babylon*; Price, *Little Science, Big Science.*

3. The documentation movement has been well studied in the history of the information sciences, but key starting points in the literature include Farkas-Conn, *From Documentation to Information Science*; Hahn and Buckland, eds., *Historical Studies in Information Science*; and Richards, *Scientific Information in Wartime.*

4. Metadata is "data about data." If you write an article about a TV show, for example, you

might "tag" it with that show's name, so that someone can click and see all other articles also about that TV show. That tag is metadata. Another common example is a hashtag on Twitter. It lets you categorize and search through your big data set.

5. Rayward, "The Origins of Information Science and the International Institute of Bibliography / International Federation for Information and Documentation (FID)," in Hahn and Buckland, *Historical Studies in Information Science.*

6. Marion, "The Library as an Adjunct to Industrial Laboratories." Citation from Williams, "The Documentation and Special Libraries Movement in the United States, 1910–1960," 174.

7. Williams, "The Documentation and Special Libraries Movement in the United States, 1910–1960," in Hahn and Buckland, *Historical Studies in Information Science.*

8. On Field and the Concilium, see Burke, *Information and Intrigue.*

9. See, for example, Gordin, *Scientific Babel,* 159–86.

10. Heath and Hetherington, *Industrial Research and Development in the United Kingdom.*

11. For a broad overview, see Graham, *Science in Russia and the Soviet Union.*

12. MacLeod, "Secrets among Friends."

13. MacLeod, "Secrets among Friends," 213n48.

14. MacLeod, "Secrets among Friends," 217.

15. MacLeod, "Secrets among Friends," 220. Citing YUL, Bumstead Papers, Bumstead to Luetta, 20 April 1918.

16. MacLeod, "Secrets among Friends," 221.

17. Meckler, *Micropublishing*; Richards, "Scientific Information for Stalin's Laboratories, 1945–1953," 14–15.

18. Meckler, *Micropublishing.* One of the practicalities overlooked by these microfilm enthusiasts, as many historians can attest, is the quality of the end-user experience relative to traditional books.

19. Richards, *Scientific Information in Wartime,* 16.

20. On the history of Aslib, see Richards, "Aslib at War," and Black, Muddiman, and Plant, *The Early Information Society.*

21. Richards, "Aslib at War," 280.

22. Richards, *Scientific Information in Wartime,* 28.

23. Cited in Muddiman, "Red Information Science," 260.

24. Richards, *Scientific Information in Wartime,* 37–38.

25. Burke, *Information and Intrigue,* 198; Farkas-Conn, *From Documentation to Information Science,* 41–43.

26. Farkas-Conn, *From Documentation to Information Science,* 89.

27. Burke, *Information and Intrigue,* 150, 198; MacDonald, "Implications of Nationalism in the Development of the Scientific Information Infrastructure in North America in the Decades Leading Up to World War II."

28. Krige, *American Hegemony and the Postwar Reconstruction of Science in Europe.*

29. Richards, *Scientific Information in Wartime,* 16–17.

30. This and the following paragraph draw upon Mikhalov et al., *Scientific Communications and Informatics.*

31. Richards, "The Movement of Scientific Knowledge from and to Germany under National Socialism," 402–3; Burleigh, *Germany Turns Eastward,* 33.

32. Graham, *Science in Russia and the Soviet Union*; Kojevnikov, *Stalin's Great Science.*

33. Wright, *Cataloging the World,* 9–10; Richards, "Aslib at War," 283.

34. Agar, *The Government Machine,* 201, 235.

35. Cortada, *All the Facts.*

36. Richards, "Aslib at War."

37. Richards, "Aslib at War," 283.

38. Richards, "Aslib at War," 284.

39. Richards, "Aslib at War," 291.

40. Farkas-Conn, *From Documentation to Information Science*, 103.

41. Richards, *Scientific Information in Wartime*, 79.

42. Richards, "Aslib at War," 288.

43. Richards, "Aslib at War," 291.

44. Richards, "Aslib at War," 291.

45. See the chapter "Wyart contre Gérard, ou la guerre des bibliographies" in Chevassus-au-Louis, *Savants sous l'Occupation*, 127–37; and Richards, "Aslib at War."

46. Gordin, *Scientific Babel*, 159–63.

47. Richards, *Scientific Information in Wartime*, 101–34.

48. Richards, *Scientific Information in Wartime*, 113.

49. Richards, *Scientific Information in Wartime*, 114.

50. "Sur la Maison de la Chimie (Affaire Gerard)," 1946. ANFF, CNRS records, RG 19780284, carton 55.

51. F. Joliot to Marston T. Bogert, 2 July 1945. ANFF, CNRS records, RG 19780284, carton 55.

52. Marston T. Bogert to Jean Gerard, 28 November 1945. ANFF, CNRS records, RG 19780284, carton 55.

53. Union Internationale de Chimie to Charles L. Parsons, American Chemical Society, 29 November 1945. ANFF, CNRS records, RG 19780284, carton 55.

54. Marston T. Bogert to Frédéric Joliot, 7 January 1946. ANFF, CNRS records, RG 19780284, carton 55; Marston T. Bogert to Frédéric Joliot, 8 January 1946. ANFF, CNRS records, RG 19780284, carton 55; Marston T. Bogert to Frédéric Joliot, 1 February 1946. ANFF, CNRS records, RG 19780284, carton 55.

55. Gimbel, *Science, Technology, and Reparations*, 64–66.

56. Gimbel, *Science, Technology, and Reparations*, 64–66.

57. L. R. Poole to G. S. Mansell, 20 December 1946. TNA: PRO BT 211/17.

58. "History: The Work in Germany is Necessarily . . . ," undated. RG 40, box 3, NACP.

59. "History: The Work in Germany is Necessarily . . . ," undated. RG 40, box 3, NACP.

60. Gordin, *Scientific Babel*, 316.

61. T. G. Haertel to Lloyd R. Worden, 22 October 1945. RG 40, box 2, NACP; Tell Berna to Joseph T. Mayer, 13 May 1946. RG 40, box 12, NACP.

62. T. G. Haertel to Lloyd R. Worden, 22 October 1945. RG 40, box 2, NACP; Tell Berna to Joseph T. Mayer, 13 May 1946. RG 40, box 12, NACP.

63. "The Chemical Problem in Germany," undated. RG 40, box 3, NACP.

64. "Record of a Meeting on Documents at FIAT," 5 February 1946. TNA: PRO BT 211/17.

65. "Copy of a Brief Handed to Major J. Day, Board of Trade, before Departure for Washington: Dissemination to Industry of Intelligence from Unclassified German Documents," undated. TNA: PRO BT 211/17.

66. "Copy of a Brief Handed to Major J. Day, Board of Trade, before Departure for Washington: Dissemination to Industry of Intelligence from Unclassified German Documents," undated. TNA: PRO BT 211/17.

67. "Memorandum for Mr. Norman," undated. TNA: PRO BT 211/17.

68. Unintelligible to Derek Wood, 10 February 1947. TNA: PRO BT 211/24.

69. L. R. Poole to George L. Powell, October 1946. TNA: PRO BT 211/17.

70. "How Easily Can We Obtain Copies of German Industrial Documents from the US Office of Technical Services? A Brief Summary Compiled from the Reports of H. W. Hirsch. (F.D.U.) from Washington during the Period 1st October 1946–1st February, 1947," undated. TNA: PRO BT 211/17.

71. "G.D. 743/46. Mr. Wood Agrees That It Is . . . ," undated. TNA: PRO BT 211/42.

72. A. King to G. S. Mansell, 16 September 1946. TNA: PRO BT 211/17. On this general British attitude, see chapter 2.

73. A. King to G. S. Mansell, 16 September 1946. TNA: PRO BT 211/17.

74. "1947 Budget Estimates: General Justification, Technical and Scientific Services," undated. RG 40, box 58, NACP.

75. L. R. Poole, "Memorandum: I Have Today Had a Long Conversation . . . ," 7 July 1948. TNA: PRO BT 211/17.

76. Krammer, "German Energy Secrets Recycled."

77. On (over)classification and science, see, for example, Galison, "Removing Knowledge"; Seidel, "Secret Scientific Communities"; Wellerstein, "Knowledge and the Bomb"; and Westwick, "Secret Science."

78. "Appendix A: Memorandum for the Communications Subcommittee, Technical Industrial Intelligence Committee," 18 August 1945. General Records, Department of Commerce, Record Group 40 (RG 40), box 23, NACP.

79. G. D. Taylor to T. G. Haertel, 4 July 1945. RG 40, box 23, NACP.

80. "Proposed Visit of Secretary, BIOS to Washington," undated. British National Archives (BNA): Public Records Office (PRO) BT 211/24.

81. H. S. Stewart-Jones, "Appendix 'A' to Minutes of the Working Party," 26 August 1946. TNA: PRO BT 211/13.

82. —— to MAM Robb, Esq., Foreign Office, 24 February 1947. TNA: PRO BT 211/161.

83. Ralph Shaw, "The Publication Board," *College and Research Libraries* (April 1946): 105.

84. Hewlett and Anderson, *A History of the United States Atomic Energy Commission, Vol. I*, 1–2, 647; Hutchinson, "The Manhattan Project Declassification Program." Citation from Westwick, "Secret Science," 367.

85. Aspray, "Command and Control, Documentation, and Library Science."

86. Aspray, "Command and Control, Documentation, and Library Science."

87. Burke, *Information and Intrigue.*

88. Burke, *Information and Intrigue*, 334–35.

89. Gimbel, *Science, Technology, and Reparations*, 70–72.

90. Coleman, "The Mass Production of Translation."

91. Coleman, "The Mass Production of Translation," 25. For more on Coleman and the scientific and technical translation industry in which he was a part, see Gordin, *Scientific Babel.*

92. Gordin, *Scientific Babel.*

93. On the AEC Technical Information Service, see Warheit, "The Library Program of the US Atomic Energy Commission Technical Information Service."

94. Williams, "The Documentation and Special Libraries Movement in the United States, 1910–1960," 174–75.

95. Farkas-Conn, *From Documentation to Information Science*, 99–119.

96. Farkas-Conn, *From Documentation to Information Science*, 99–119.

97. Farkas-Conn, *From Documentation to Information Science*, 106.

98. Krementsov, *Stalinist Science*, 116.

99. Gordin, *Scientific Babel*, 279–89.

100. Giliarevskii, "Soviet Scientific and Technical Information System"; Richards, "The Soviet Overseas Information Empire and the Implications of Its Disintegration."

101. Muddiman, "Red Information Science."

102. Richards, *Scientific Information in Wartime*, 130.

103. Houghton, *Out of the Dinosaurs*, 17. Citation from Till, "Predecessors of Preprint Servers," 131.

104. Richards, *Scientific Information in Wartime*, 125–35.

105. Richards, *Scientific Information in Wartime*, 128.

106. Richards, *Scientific Information in Wartime*, 126.

107. Bush, *Science, the Endless Frontier*, 19. On this work's political impact, see Kevles, "The National Science Foundation and the Debate over Postwar Research Policy, 1942–1945."

Chapter 7 · *Legacies of Intellectual Reparations Programs*

1. Examples include survey results in Behrman, "Licensing Abroad under Patents, Trademarks, and Know-How by US Companies"; Behrman and Schmidt, "New Data on Foreign Licensing"; Schmidt, "Licensing Know-How, Patents, and Trademarks Abroad"; Siegel, "Scientific Discovery, Invention, and the Cultural Environment"; Bleeke and Rahl, "The Value of Territorial and Field-of-Use Restrictions in the International Licensing of Unpatented Know-How."

2. "Protection of Know-How, Executive Committee of the ICC, 21 February 1961." In International Chamber of Commerce, "Statements and Resolutions of the ICC, 1959–1961," in *XVIIIth Congress of the ICC*.

3. Zarem, "Know-How, Our Maginot Line."

4. On tacit knowledge in general, see, for example, Collins, *Tacit and Explicit Knowledge*; Kaiser, *Drawing Theories Apart*; Polanyi, *Personal Knowledge*; and Polanyi, *The Tacit Dimension*.

5. Joint Intelligence Committee, "Acquisition of Germany Technical Information of an Industrial Nature," 12 October 1944. RG 40, Entry 75, box 62, NACP, 5.

6. "TIIC/C 8th Meeting," 26 April 1945. RG 40, Entry 75, box 40, NACP, 7.

7. Lucius D. Clay to Quadripartite Coordinating Committee, "Removal from Germany of Documents, Equipment, and Information for Technical Purposes," December 1945. RG 260, box 17/03, folder 13, NACP.

8. "Facilities for Technical Investigations in Germany by Individual British Firms," 22 October 1946. TNA: PRO BT 211/21.

9. "Business Bulletin: A Special Background Report on Trends in Industry and Finance," *Wall Street Journal*, September 7, 1944.

10. David Bartlett, "Engineering."

11. Zarem, "Know-How, Our Maginot Line"; "Foreign Market for US Know-How Greater than Ever"; Tolin, "United States Shares Chemical Know-How with the Other Americas"; "Capitalizing on Foreign Know-How."

12. "Capitalizing on Foreign Know-How."

13. *Hansard*, HL, 29 April 1948, Vol. 155, col. 591–97.

14. *Hansard*, HL, 8 October 1946, Vol. 427, col. 46–47; *Hansard*, HC, 12 November 1956, Vol. 560, col. 620–21.

15. "Protection of Know-How, Executive Committee of the ICC, 21 February 1961." In International Chamber of Commerce, "Statements and Resolutions of the ICC, 1959–1961." For a discussion of the history and organization of the ICC, see Kelly, "The International Chamber of Commerce."

16. Ladas, "Legal Protection of Know-How."

17. Wolfe, "Restrictions in Know-How Agreements."

18. Macdonald, "Know-How Licensing and the Antitrust Laws," 252.

19. Behrman and Schmidt, "New Data on Foreign Licensing."

20. John D. Leonard, "Know-How for Sale: British Offer US Firms Their Industrial Secrets in Exchange for Dollars," *Wall Street Journal*, 7 October 1947. By the 1970s, another article under the same title, "Know-How for Sale," would later warn that "Concern Grows over Rising US Exports of Skilled Technology to Overseas Firms."

21. Stokes, *Opting for Oil*, 115.

22. Stokes, *Opting for Oil*, 115.

23. Stokes, *Opting for Oil*, 118.

24. Stokes, *Opting for Oil*, 118. This quote is from "Bericht über Studienreise der Herren Dr. Hagedorn/Dr. Weinbrenner nach den USA in der Zeit von 12.9. bis 8.11 1952," BWA, 700/453.1.

25. This discussion builds primarily from Erker, "The Long Shadow of Americanization," in Zeitlin and Herrigel, *Americanization and Its Limits*.

26. W. C. Bryant, "Synthetic Secrets: Rubber Firms Ask US to Alter Their War Pact Pooling All Discoveries," *Wall Street Journal*, August 12, 1947.

27. BIOS Overall Report 7, "The Rubber Industry in Germany during the Period 1939–45," 1948, I.

28. CA, letter, 24 January 1947, 6525 I/56, A25. Citation from Erker, "The Long Shadow of Americanization," 301.

29. CA, letter, 24 January 1947, 6525 I/56, A25. Citation from Erker, "The Long Shadow of Americanization," 302.

30. Werner Abelshauser, *German Industry and Global Enterprise*, 514–17.

31. Abelshauser, *German Industry and Global Enterprise*, 354.

32. Minutes of the managing board meeting of March 2, 1964; BASF RA. Citation from Abelshauser, *German Industry and Global Enterprise*, 519.

33. Stokes, *Opting for Oil*, 6.

34. Stokes, *Opting for Oil*, 75.

35. For more on this topic, see O'Reagan, "Know-How in Postwar Business and the Law."

36. Bartlett, "Engineering."

37. Miles, "The Transfer of Technical Know-How"; Beach, "A Question of Property Rights; Schmidt, "Licensing Know-How, Patents, and Trademarks Abroad."

38. Brewster, *Antitrust and American Business Abroad*, 159.

39. Creed and Bangs, " 'Know-How' Licensing and Capital Gains," 93–94.

40. Stedman, "Legal Problems in the International and Domestic Licensing of Know-How," 250–51.

41. Zarem, "Know-How, Our Maginot Line."

42. Carew, "The Anglo-American Council on Productivity (1948–52)," 52. On the Anglo-American Council on Productivity and general British interest in American industrial technology, see also Carew, *Labour under the Marshall Plan*; McGlade, "Americanization," in Zeitlin and Herrigel, *Americanization and Its Limits*; Tiratsoo and Gourvish, " 'Making It Like Detroit' "; and Tiratsoo and Tomlinson, "Exporting the 'Gospel of Productivity.' "

43. Carew, "The Anglo-American Council on Productivity (1948–52)," 53–55.

44. Carew, "The Anglo-American Council on Productivity (1948–52)," 55.

45. L. Rostas, "Industrial Production, Productivity and Distribution in Britain, Germany and the United States," *Economic Journal* 53 (April 1943): 39–54. These numbers vary slightly from those given in Tiratsoo and Tomlinson, "Exporting the 'Gospel of Productivity.' " They claim 66 teams of 956 participants, citing: Francis E. Rogers, "Report of the United Kingdom Technical Exchange and Section 115-K Program," 6 September 1956, 1–8, box 5, file on "U.K. Productivity—General (Directors' Files) 1955–7," Subject Files of the Director 1953–7, Office of the Director, Mission to the U.K., Records Group 469, National Archives, Washington, DC (hereafter Rogers); and *EPA Information Bulletin*, no. 6–7 (December 1954–January 1955): 10.

46. Carew, "The Anglo-American Council on Productivity (1948–52)," 59.

47. Tiratsoo and Tomlinson, "Exporting the 'Gospel of Productivity,' " 48–49. They cite this to Francis E. Rogers to William L. Batt, 5 August 1952, box 6, file on "U.K. Productivity. 115K Agreement . . . ," Subject Files of the Director 1953–7, Office of the Director, Mission to the U.K., Records Group 469, National Archives, Washington, DC.

48. British Productivity Council, A *Review of Productivity in the Bronze and Brass Casting*

Industry (London, 1955), 7. Citation from Tiratsoo and Tomlinson, "Exporting the 'Gospel of Productivity,'" 53.

49. G. Macrea, "New Ideas Enliven an Old Industry," *Business* 85 (1955): 107–111. Citation from Tiratsoo and Tomlinson, "Exporting the 'Gospel of Productivity,'" 62.

50. Behrman, "Licensing Abroad under Patents, Trademarks, and Know-How by US Companies," 191.

51. *Economist*, Foreign Report, Economist Intelligence Unit, 3 February 1949. Citation from Carew, "The Anglo-American Council on Productivity (1948–52)," 55.

52. Gillingham, *Coal, Steel, and the Rebirth of Europe, 1945–1955*; Hitchcock, *France Restored*; Kiersch, "Die französische Deutschlandpolitik 1945–1949," in Scharf and Schröder, *Die Deutschlandpolitik Frankreichs und die Französische Zone, 1945–1949*.

53. Kipping, "A Slow and Difficult Process," in Zeitlin and Herrigel, *Americanization and Its Limits*.

54. International Cooperation Administration (ICA), *European Productivity and Technical Assistance Programs: A Summing Up (1948–58)*. Citation from Tiratsoo and Tomlinson, "Exporting the 'Gospel of Productivity,'" 209.

55. Tiratsoo and Tomlinson, "Exporting the 'Gospel of Productivity,'" 211.

56. In terms of academic science, see Krige, *American Hegemony and the Postwar Reconstruction of Science in Europe*.

57. Krige, "Atoms for Peace, Scientific Internationalism, and Scientific Intelligence," 180; Hewlett and Holl, *Atoms for Peace and War, 1953–1961*, 324; Kohlrausch and Trischler, *Building Europe on Expertise: Innovators, Organizers, Networkers*.

58. Stokes, *Constructing Socialism*, 89–92.

59. Stokes, *Constructing Socialism*, 91.

60. Stokes, *Constructing Socialism*, 92.

61. Most of this discussion draws from Zhang, Zhang, and Yao, "Technology Transfer from the Soviet Union to the People's Republic of China." See also Zuoyue Wang, "The Chinese Developmental State during the Cold War"; Lewis and Xue, *China Builds the Bomb*.

62. Mao, Zai Chengdu huiyi shang de liuci jianghua [Six speeches at the Chengdu meeting, on 10 March 1958], in *Mao Zedong Sixiang Wansui* [Long live Mao Zedong thought, 1958–1959], 27–29. Citation from Zhang, Zhang, and Yao, "Technology Transfer from the Soviet Union to the People's Republic of China," 113.

63. Winter, "The Licensing of Know-How to the Soviet Union," 172.

64. "It's Still 'No' to Soviets."

65. "ICI Changes Its Soviet Trade Policy."

66. Winter, "The Licensing of Know-How to the Soviet Union," 172.

67. On Cold War propaganda, one excellent source is Cull, *The Cold War and the United States Information Agency*. However, Cull make no real mention of the role of science and technology in Cold War propaganda, a literature that is expanding rapidly. See, for example, Cullather, *The Hungry World*; Doel and Harper, "Prometheus Unleashed"; Krige, "Technological Leadership and American Soft Power"; Krige, Callahan, and Maharaj, *NASA in the World*.

68. On modernization theory, see Ekbladh, *The Great American Mission*; Gilman, *Mandarins of the Future*; and Latham, *The Right Kind of Revolution*.

69. "Truman Inaugural Address, January 20, 1949," last modified January 20, 1949, https://www.trumanlibrary.org/whistlestop/50yr_archive/inagural20jan1949.htm.

70. Truman, "Special Message to the Congress on the Mutual Security Program," in *Public Papers of the Presidents of the United States*.

71. International Technical Cooperation Act of 1949 ("Point IV" Program): Hearings before the Committee on Foreign Affairs, House of Representatives, 81st Congress, 1st Session (1949), https://catalog.hathitrust.org/Record/100665946, 6.

72. Ekbladh, *The Great American Mission*, 174.

73. "Possible Questions and Suggested Answers concerning the President's Technical Assistance Proposal," April 12, 1949. Truman Papers, Official File. OF 426: Foreign Relief. https://www.trumanlibrary.org/whistlestop/study_collections/pointfourprogram/documents/index.php?documentid=4-7&pagenumber=1.

74. United Nations General Assembly Resolution 200(III). As cited in Mehos and Moon, "The Uses of Portability," in Hecht, *Entangled Geographies*, 57.

75. Sharp, "The Institutional Framework for Technical Assistance," 343.

76. Trager and Trager, "Exporting and Training Experts," 93.

77. Trager and Trager, "Exporting and Training Experts," 93. As cited in Mehos and Moon, "The Uses of Portability."

78. Goekjian, "Legal Problems of Transferring Technology to the Third World." On developing nations' efforts to address know-how in the law, see also Ladas, "Legal Protection of Know-How"; Wolfe, "Restrictions in Know-How Agreements"; Correa, "Legal Nature and Contractual Conditions in Know-How Transactions"; Bleeke and Rahl, "The Value of Territorial and Field-of-Use Restrictions in the International Licensing of Unpatented Know-How"; Suzanne F. Greenberg, "The WIPO Model Laws for the Protection of Unpatented Know-How."

79. On this point, see Biagioli, Jaszi, and Woodmansee, eds., *Making and Unmaking Intellectual Property*; Khan, *The Democratization of Invention*; May and Sell, *Intellectual Property Rights*; Sherman and Bently, *The Making of Modern Intellectual Property Law*; Wells, *Antitrust and the Formation of the Postwar World*.

80. Ladas, "Legal Protection of Know-How."

81. Bleeke and Rahl, "The Value of Territorial and Field-of-Use Restrictions in the International Licensing of Unpatented Know-How."

82. Bleeke and Rahl, "The Value of Territorial and Field-of-Use Restrictions in the International Licensing of Unpatented Know-How."

83. N. D. Crane and C. S. Weaver to Ralph Osborne, "Exploitation of Heidenheim Personnel—FIAT Technical Institute." 1 March 1946. Records of the US Occupation Headquarters, World War II, Record Group 260 (RG 260), FIAT Administrative Records 1945–1947, box 17/11, folder 1, NACP.

84. See John Gimbel, *Science, Technology, and Reparations*, 37–40.

85. Pash, *The Alsos Mission*; Goudsmit, *Alsos*.

86. J. J Macfarland to Asst. Chief of Staff, G-2, "Exploitation and Denial of German Scientist—Dr. RICHTER," 5 September 1946. RG 260.3.6, box 17/1, folder 35, NACP.

87. F. J. Broomfield to Derek Wood, 8 January 1947. TNA: PRO BT 211/66.

88. Mastanduno, *Economic Containment*, 69.

89. "Potomac Postscripts."

90. As cited in Yergin, *Shattered Peace*, 92–93.

91. Truman was quoted by Harold Stassen in US Congress, Senate, Committee on Government Operations, Permanent Subcommittee on Investigations, *East-West Trade*, hearings, 84th Congress, 2nd Session, February 15–17, 20, and March 6, 1956, 450. Citation from Mastanduno, *Economic Containment*, 93.

92. Haynes, Klehr, Vassiliev, Redko, and Shabad, *Spies*; Haynes and Klehr, *Venona*; Weinstein and Vassiliev, *The Haunted Wood*; Klehr, Haynes, and Anderson, *The Soviet World of American Communism*.

93. Haynes et al., *Spies*; Weinstein and Vassiliev, *The Haunted Wood*; Haynes and Klehr, *Venona*.

94. Maurice C. Ernst, "Economic Intelligence in CIA," in Westerfield, *Inside CIA's Private World*.

95. An accessible starting point here is Wolfe, *Competing with the Soviets*. More advanced discussions include Gordin, *Red Cloud at Dawn*; Rhodes, *The Making of the Atomic Bomb*.

96. Felix Belair, "US Will Not Share Atom Bomb Secret, President Asserts: Calls Industrial 'Know How' the Most Important Factor, Not Scientific Knowledge," *New York Times*, 9 October 1945.

97. Winston Churchill, address to Parliament, 7 November 1945. Citation from Kaiser, "The Atomic Secret in Red Hands?," 35.

98. *Hansard*, HC, 8 October 1946, Vol. 427, col. 43–98. http://hansard.millbanksystems.com /commons/1946/oct/08/atomic-energy-bill.

99. Kaiser, "The Atomic Secret in Red Hands?," 34.

100. Wang, *American Science in an Age of Anxiety*.

101. Galison, "Removing Knowledge"; Westwick, "Secret Science."

102. Chang, *Thread of the Silkworm*.

103. On Atoms for Peace, see Chernus, *Eisenhower's Atoms for Peace*; Hewlett and Holl, *Atoms for Peace and War, 1953–1961*; Krige, "Atoms for Peace, Scientific Internationalism, and Scientific Intelligence."

104. Dwight D. Eisenhower, "Address by Dwight D. Eisenhower on Psychological Warfare, October 8, 1952," as cited in Medhurst, "Atoms for Peace and Nuclear Hegemony," 572.

105. NSC, "Peaceful Uses of Atomic Energy," NSC-5507/2, 12 March 1955. https://history .state.gov/historicaldocuments/frus1955–57v20/d14.

106. On these programs, see Abraham, *The Making of the Indian Atomic Bomb*, and Cohen, *Israel and the Bomb*. Of course, each nation's path toward (or away from) nuclear weapons depended far more on other factors (domestic politics, scientific infrastructure, etc.) than it did on American aid.

107. Schmid, "Nuclear Colonization?," in Hecht, *Entangled Geographies*, 133.

108. Josephson, *Red Atom*, 174.

109. Josephson, *Red Atom*, 174; Krige, "Atoms for Peace, Scientific Internationalism, and Scientific Intelligence."

110. Schmid, "Nuclear Colonization?," 131.

111. Modelski, *Atomic Energy in the Communist Bloc*, 157–61; Albrecht, Heinemann-Grüder, and Wellmann, *Die Spezialisten*.

112. Zeitlin and Herrigel, eds., *Americanization and Its Limits*, 6. For a broader take on the idea of "Americanization" of Europe, see De Grazia, *Irresistible Empire*. For an excellent recent summary of historical thought on the term and idea, see Berghahn, "The Debate on 'Americanization' among Economic and Cultural Historians."

Conclusion

1. For a list of these reports, see US Department of Commerce, Office of Technical Services et al., "FIAT Reports." A more detailed listing is in the Bibliography of Scientific and Industrial Reports, published by the US Department of Commerce's Office of Technical Services, 1946–49: https://catalog.hathitrust.org/Record/009487225.

2. On German economic history, see, for example, Abelshauser, *Deutsche Wirtschaftsgeschichte von 1945 bis zur Gegenwart*; Berghahn, ed., *Quest for Economic Empire*.

3. David L. Morton, "'The Rusty Ribbon,'" 589.

4. Lloyd Worden, "Memo to John Green, Report on Trip to Wright Field," 20 August 1946. General Records, Department of Commerce, RG 40, Box 3, NACP.

5. Trevor Evans, "Smith and the Secrets of Schmidt," *Daily Express*, 9 October 1945.

6. Examples of key historical works on technology transfer include articles in the now-defunct journal *Comparative Technology Transfer and Society* (2003–2009), available at https://

muse.jhu.edu/journal/237. Excellent book-length studies emphasizing the difficulty of international technology transfer include Alder, *Engineering the Revolution*; Harris, *Industrial Espionage and Technology Transfer*; Macrakis, *Seduced by Secrets*; and Palmer, *Dictatorship of the Air*.

7. Milward, *The Reconstruction of Western Europe, 1945–51*, 167. Citation from Stokes, *Divide and Prosper*, 86.

8. Stokes, *Divide and Prosper*, 86–106.

9. Edgerton, *Science, Technology and the British Industrial "Decline", 1870–1970*; Edgerton, *Warfare State*.

10. This American figure is from Gimbel, *Science, Technology, and Reparations*, 79. British numbers are from Glatt, "Reparations and the Transfer of Scientific and Industrial Technology from Germany," 163.

11. On America's economic situation emerging from the war, see Lind, *Land of Promise*.

12. Gordin, *Red Cloud at Dawn*; Holloway, *Stalin and the Bomb*; Richelson, *Spying on the Bomb*.

13. Rivas, *Missionary Capitalist*.

14. Naimark, *The Russians in Germany*, 223.

15. For a summary of many of these, see Lobel, *Talent Wants to Be Free*.

Primary Sources

Archives Nationale, Fontainebleau
Archives Nationale, Paris
Bibliothéque Nationale, Paris
British Library
British National Archives, Kew
The National Archives at College Park, Maryland

Bartlett, David. "Engineering—An Invisible Export." *Electrical Engineering* 68, no. 5 (May 1949): 379–82.

Browne, C. A. "The Role of Refugees in the History of American Science." *Science* 91, no. 2357 (1940): 203–8.

"Capitalizing on Foreign Know-How: Importing or Exporting a Foreign Process and Know How Involves Much More Than a Licensing Agreement." *Chemical and Engineering News* 34, no. 28 (1956): 3352.

"The CIA and the Nazis." *Conspiracy?* The History Channel, December 19, 2004.

Coleman, Earl Maxwell. "The Mass Production of Translation—For a Limited Market." *Publishing Research Quarterly* 10, no. 4 (1995): 22–29.

Faragher, W. F. "Collecting German Industrial Information." *Chemical and Engineering News* 26, no. 52 (December 27, 1948): 3816–20.

"Foreign Market for US Know-How Greater Than Ever." *Chemical and Engineering News* 29, no. 45 (1951): 4674–76.

Hughes, Francis. "Notes on the Prospect Confronting Post-War British Patent Property." *Journal of the Patent Office Society* 27, no. 11 (1945): 729–69.

Hutchinson, W. S. "The Manhattan Project Declassification Program." *Bulletin of the Atomic Scientists* 2, nos. 9–10 (1946): 14–15.

"ICI Changes Its Soviet Trade Policy: Company is Prepared to Supply Plant and Know-How in a Package Deal; Soviets Favor New Policy." *Chemical and Engineering News* 42, no. 1 (1964): 42.

International Chamber of Commerce. "Statements and Resolutions of the ICC, 1959–1961; Copenhagen 22–27 May 1961." Paris: International Chamber of Commerce, 1961.

"It's Still 'No' to Soviets: Chemical Industry Spokesmen Again Slam Door on Red Angling for Technical Know-How." *Chemical and Engineering News* 37, no. 16 (1959): 28.

Kirkpatrick, Sidney. "The Clock Strikes Twelve." *Chemical and Metallurgical Engineering* 43 (May 1945): 93.

Macrea, G. "New Ideas Enliven an Old Industry." *Business* 85 (1955): 107–11.

Major, Randolph T. "Chemistry in Postwar Europe." *Science* 111, no. 2884 (1950): 366–67.

Marion, G. E. "The Library as an Adjunct to Industrial Laboratories." *Library Journal* 35, no. 9 (1910): 400–404.

Miles, Arnold. "The Transfer of Technical Know-How." *Public Administration Review* 13, no. 1 (1953): 50–53.

Nettl, Peter. "German Reparations in the Soviet Empire." *Foreign Affairs*, January 1951. https://www.foreignaffairs.com/articles/germany/1951-01-01/german-reparations-soviet-empire.

"Potomac Postscripts: Office of International Trade Initiates Voluntary Program to Control Export of Unclassified Technical Information and Know-How." *Chemical and Engineering News* 27, no. 47 (1949): 3469–3470.

Rostas, L. "Industrial Production, Productivity and Distribution in Britain, Germany and the United States." *Economic Journal* 53 (April 1943): 39–54.

Shaw, Ralph. "The Publication Board." *College and Research Libraries* (April 1946): 105–8.

Spencer, Richard. "The German Patent Office." *Journal of the Patent Office Society* 31, no. 2 (1949): 79–87.

Tolin, Harry Wright. "United States Shares Chemical Know-How with the Other Americas." *Chemical and Engineering News* 23, no. 16 (1945): 1436–38.

"Truman's Inaugural Address, January 20, 1949." Last modified January 20, 1949. Accessed June 24, 2016. https://www.trumanlibrary.org/whistlestop/50yr_archive/inagural20jan1949.htm.

US Department of Commerce, Office of Technical Services, Arlene Blackburn, W. Lowry, Inez Wallace, and Robert Bolin. "FIAT Reports: A Bibliography and Index of Reports Resulting from American Investigations of German Industry." *Electronic Reference Materials* (January 1, 1948). http://digitalcommons.unl.edu/libelecrefmat/6.

Vaughan, Floyd L. "Recent Patent Legislation in Great Britain." *Journal of the Patent Office Society* 32, no. 9 (1950): 692–707.

"We'll Just Never Learn." *Aero Digest* (April 1946): 169–70.

Zarem, A. M. "Know-How, Our Maginot Line." *Chemical and Engineering News* 30, no. 45 (1952): 4718–19.

Secondary Sources
Books and Dissertations

Abelshauser, Werner. *Deutsche Wirtschaftsgeschichte von 1945 bis zur Gegenwart*. München: Beck, 2011.

———. *The Dynamics of German Industry: Germany's Path toward the New Economy and the American Challenge*. New York: Berghahn Books, 2005.

———. *German Industry and Global Enterprise: BASF: The History of a Company*. Cambridge: Cambridge University Press, 2004.

Abraham, Itty. *The Making of the Indian Atomic Bomb: Science, Secrecy and the Postcolonial State*. New Delhi: Orient Longman, 1999.

Agar, Jon. *The Government Machine: A Revolutionary History of the Computer*. Cambridge, MA: MIT Press, 2003.

Albrecht, Ulrich, Andreas Heinemann-Grüder, and Arend Wellmann. *Die Spezialisten: Deutsche Naturwissenschaftler und Techniker in der Sowjetunion nach 1945*. Berlin: Dietz, 1992.

Albrecht, Ulrich, and Randolph Nikutta. *Die Sowjetische Rüstungsindustrie*. Köln: Weltdeutscher Verlag, 1989.

Alder, Ken. *Engineering the Revolution: Arms and Enlightenment in France, 1763–1815*. Chicago: University of Chicago Press, 2010.

Alter, Peter. *The Reluctant Patron: Science and the State in Britain, 1850–1920*. New York: Berg, 1987.

Ash, Mitchell G., and Alfons Söllner. *Forced Migration and Scientific Change: Emigre German-*

Speaking Scientists and Scholars after 1933. Washington, DC: German Historical Institute, 1996.

Augustine, Dolores L. *Red Prometheus: Engineering and Dictatorship in East Germany, 1945–1990*. Cambridge, MA: MIT Press, 2007.

Backer, John. *Priming the German Economy: American Occupational Policies, 1945–1948*. Durham, NC: Duke University Press, 1971.

Balabkins, Nicholas. *Germany under Direct Controls: Economic Aspects of Industrial Disarmament, 1945–1948*. New Brunswick, NJ: Rutgers University Press, 1964.

Bar-Zohar, Michael. *The Hunt for German Scientists*. London: Barker, 1967.

Bauman, Zygmunt. *Modernity and the Holocaust*. Ithaca, NY: Cornell University Press, 1989.

Beissinger, Mark R. *Scientific Management, Socialist Discipline, and Soviet Power*. Cambridge, MA: Harvard University Press, 1988.

Bellamy, Matthew J. *Profiting the Crown: Canada's Polymer Corporation, 1942–1990*. McGill-Queen's Press, 2005.

Berghahn, Volker R. *The Americanisation of West German Industry: 1945–1973*. New York: Berg, 1986.

———, ed. *Quest for Economic Empire: European Strategies of German Big Business in the Twentieth Century*. New York: Berghahn Books, 1996.

Biagioli, Mario, Peter Jaszi, and Martha Woodmansee, eds. *Making and Unmaking Intellectual Property: Creative Production in Legal and Cultural Perspective*. Chicago: University of Chicago Press, 2011.

Biddle, Wayne. *Dark Side of the Moon: Wernher Von Braun, the Third Reich, and the Space Race*. New York: W. W. Norton, 2009.

Black, Alistair, Dave Muddiman, and Helen Plant. *The Early Information Society: Information Management in Britain before the Computer*. Brookfield, VT: Ashgate, 2007.

Boehm, Klaus. *The British Patent System: I. Administration*. Cambridge: Cambridge University Press, 1967.

Borkin, Joseph, and Charles A. Welsh. *Germany's Master Plan: The Story of Industrial Offensive*. New York: Duell, Sloan and Pearce, 1943.

Bower, Tom. *The Paperclip Conspiracy: The Hunt for the Nazi Scientists*. Boston: Little, Brown, 1987.

Bowler, Peter J. *Science for All: The Popularization of Science in Early Twentieth-Century Britain*. 1st ed. Chicago: University of Chicago Press, 2009.

Boyers, Robert. *The Legacy of the German Refugee Intellectuals*. New York: Schocken Books, 1972.

Braun, Hans-Joachim. *The German Economy in the Twentieth Century: The German Reich and the Federal Republic*. New York: Routledge, 1990.

Brewster, Kingman. *Antitrust and American Business Abroad*. New York: McGraw-Hill, 1958.

British Library Board. *British Patents of Invention, 1617–1977: A Guide for Researchers*. London: British Library, 1999.

British Productivity Council. *A Review of Productivity in the Bronze and Brass Casting Industry*. London: British Productivity Council, 1955.

Brown, Mark B. *Science in Democracy: Expertise, Institutions, and Representation*. Cambridge, MA: MIT Press, 2009.

Buchheim, Christoph. *Die Wiedereingliederung Westdeutschlands in Die Weltwirtschaft 1945–1958*. München: R. Oldenbourg Verlag, 1990.

Burke, Colin. *Information and Intrigue: From Index Cards to Dewey Decimals to Alger Hiss*. Cambridge, MA: MIT Press, 2014.

———. *Information and Secrecy: Vannevar Bush, Ultra, and the Other Memex*. Metuchen, NJ: Scarecrow Press, 1994.

Burleigh, Michael. *Germany Turns Eastward: A Study of Ostforschung in the Third Reich*. Cambridge: Cambridge University Press, 1988.

Bush, Vannevar. *Science, the Endless Frontier*. Washington, DC: Government Printing Office, 1945.

Cairncross, Alec. *The Price of War: British Policy on German Reparations, 1941–1949*. New York: Blackwell, 1986.

Carew, Anthony. *Labour under the Marshall Plan: The Politics of Productivity and the Marketing of Management Science*. Detroit: Wayne State University Press, 1987.

Carson, Cathryn. *Heisenberg in the Atomic Age: Science and the Public Sphere*. Cambridge: Cambridge University Press, 2010.

Chang, Iris. *Thread of the Silkworm*. New York: Basic Books, 1995.

Chernus, Ira. *Eisenhower's Atoms for Peace*. College Station: Texas A&M University Press, 2002.

Chevassus-au-Louis, Nicholas. *Savants Sous l'Occupation: Enquête Sur La Vie Scientifique Française Entre 1940 et 1944*. Paris: SEUIL, 2004.

Cohen, Avner. *Israel and the Bomb*. New York: Columbia University Press, 2003.

Collins, Harry. *Tacit and Explicit Knowledge*. Chicago: University of Chicago Press, 2010.

Cortada, James W. *All the Facts: A History of Information in the United States since 1870*. Oxford: Oxford University Press, 2016.

Coser, Lewis A. *Refugee Scholars in America: Their Impact and Their Experiences*. New Haven, CT: Yale University Press, 1984.

Craig, Gordon Alexander, and Francis L. Loewenheim. *The Diplomats, 1939–1979*. Princeton, NJ: Princeton University Press, 1994.

Creed, John F., and Robert B. Bangs. " 'Know-How' Licensing and Capital Gains." *Patent, Trademark, Copyright Journal of Research and Education* 4 (1960): 93–107.

Crim, Brian E. *Our Germans: Project Paperclip and the National Security State*. Baltimore: Johns Hopkins University Press, 2017.

Cull, Nicholas John. *The Cold War and the United States Information Agency: American Propaganda and Public Diplomacy, 1945–1989*. Cambridge: Cambridge University Press, 2008.

Cullather, Nick. *The Hungry World: America's Cold War Battle against Poverty in Asia*. Cambridge, MA: Harvard University Press, 2010.

Daston, Lorraine J., and Peter Galison. *Objectivity*. New York: Zone Books, 2010.

Defrance, Corine. *Les Alliés Occidentaux et Les Universités Allemandes, 1945–1949*. Paris: CNRS Éditions, 2000.

Defrance, Corine, Michael Kißener, and Pia Nordblom, eds. *Wege der Verständigung Zwischen Deutschen und Franzosen Nach 1945—Zivilgesellschaftliche Annäherungen*. Tübingen: Narr Verlag, 2010.

De Grazia, Victoria. *Irresistible Empire: America's Advance through Twentieth-Century Europe*. Cambridge, MA: The Belknap Press of Harvard University Press, 2005.

Deighton, Anne. *The Impossible Peace: Britain, the Division of Germany and the Origins of the Cold War*. Oxford: Oxford University Press, 1990.

Edgerton, David. *Science, Technology and the British Industrial "Decline", 1870–1970*. Cambridge: Cambridge University Press, 1996.

———. *Warfare State: Britain, 1920–1970*. Cambridge: Cambridge University Press, 2005.

Ekbladh, David. *The Great American Mission: Modernization and the Construction of an American World Order*. Princeton, NJ: Princeton University Press, 2010.

Epstein, Katherine C. *Torpedo: Inventing the Military-Industrial Complex in the United States and Great Britain*. Cambridge, MA: Harvard University Press, 2014.

Farkas-Conn, Irene. *From Documentation to Information Science: The Beginnings and Early Development of the American Documentation Institute—American Society for Information Science*. Westport, CT: Greenwood Press, 1990.

Fermi, Laura. *Illustrious Immigrants: The Intellectual Migration from Europe, 1930–41*. Chicago: University of Chicago Press, 1968.

Fleming, Donald, and Bernard Bailyn. *The Intellectual Migration: Europe and America, 1930–1960*. Cambridge, MA: The Belknap Press of Harvard University Press, 1969.

Foschepoth, Josef, and Rolf Steininger, eds. *Die Britische Deutschland- und Besatzungspolitik 1945–1949*. Paderborn: F. Schöningh, 1985.

Gaddis, John. *The United States and the Origins of the Cold War: 1943–1946*. Ann Arbor, MI: UMI Dissertation Services, 2003.

Gemelli, Giuliana. *The "Unacceptables": American Foundations and Refugee Scholars between the Two Wars and After*. Brussels: PIE Peter Lang, 2000.

Giffard, Hermione. *Making Jet Engines in World War II: Britain, Germany, and the United States*. Chicago: University of Chicago Press, 2016.

Gillingham, John. *Coal, Steel, and the Rebirth of Europe, 1945–1955: The Germans and French from Ruhr Conflict to Economic Community*. Cambridge: Cambridge University Press, 1991.

Gilman, Nils. *Mandarins of the Future: Modernization Theory in Cold War America*. Baltimore: Johns Hopkins University Press, 2003.

Gimbel, John. *The Origins of the Marshall Plan*. Stanford, CA: Stanford University Press, 1976.

———. *Science, Technology, and Reparations: Exploitation and Plunder in Postwar Germany*. Stanford, CA: Stanford University Press, 1990.

Gispen, Kees. *Poems in Steel: National Socialism and the Politics of Inventing from Weimar to Bonn*. New York: Berghahn Books, 2002.

Glatt, Carl. "Reparations and the Transfer of Scientific and Industrial Technology from Germany: A Case Study of the Roots of British Industrial Policy and of Aspects of British Occupation Policy in Germany between Post–World War II, Reconstruction and the Korean War, 1943–1951." PhD dissertation, European University Institute, 1994.

Gordin, Michael D. *Red Cloud at Dawn: Truman, Stalin, and the End of the Atomic Monopoly*. New York: Farrar, Straus and Giroux, 2009.

———. *Scientific Babel: How Science Was Done before and after Global English*. Chicago: University of Chicago Press, 2015.

Goudsmit, Samuel. *Alsos*. Los Angeles: Tomash Publishers, 1983.

Graham, Loren R. *Science in Russia and the Soviet Union: A Short History*. Cambridge: Cambridge University Press, 1993.

Grüttner, Michael, ed. *Gebrochene Wissenschaftskulturen: Universität und Politik im 20. Jahrhundert*. Göttingen: Vandenhoeck and Ruprecht, 2010.

Hahn, Trudi Bellardo, and Michael Buckland, eds. *Historical Studies in Information Science*. Medford, NJ: American Society for Information Science, 1998.

Harris, J. R. *Industrial Espionage and Technology Transfer: Britain and France in the Eighteenth Century*. Brookfield, VT: Ashgate, 1998.

Hartcup, Guy. *The War of Invention: Scientific Developments, 1914–18*. London: Brassey's Defence Publishers, 1988.

Hayes, Peter. *Industry and Ideology: IG Farben in the Nazi Era*. Cambridge: Cambridge University Press, 1987.

Haynes, John Earl, and Harvey Klehr. *Venona: Decoding Soviet Espionage in America*. New Haven, CT: Yale University Press, 2000.

Haynes, John Earl, Harvey Klehr, Alexander Vassiliev, Philip Redko, and Steven Shabad. *Spies: The Rise and Fall of the KGB in America*. New Haven, CT: Yale University Press, 2009.

Heath, H. F., and A. L. Hetherington. *Industrial Research and Development in the United Kingdom: A Survey*. London: Faber and Faber Limited, 1946.

Hecht, Gabrielle. *Entangled Geographies: Empire and Technopolitics in the Global Cold War*. Cambridge, MA: MIT Press, 2011.

———. *The Radiance of France: Nuclear Power and National Identity after World War II.* Cambridge, MA: MIT Press, 1998.

Heer, Hannes, and Klaus Naumann, eds. *War of Extermination: The German Military in World War II.* New York: Berghahn Books, 2004.

Heilbron, John. *The Dilemmas of an Upright Man: Max Planck and the Fortunes of German Science.* Cambridge, MA: Harvard University Press, 2000.

Heilbut, Anthony. *Exiled in Paradise: German Refugee Artists and Intellectuals in America, from the 1930s to the Present.* New York: Viking, 1983.

Heim, Susanne, Carola Sachse, and Mark Walker. *The Kaiser Wilhelm Society under National Socialism.* Cambridge: Cambridge University Press, 2009.

Herrmann, Manfred. *Project Paperclip: Deutsche Wissenschaftler in diensten der U.S. Streitkräfte nach 1945.* PhD dissertation, Friedrich-Alexander-Universität Erlangen-Nürnberg, 1999.

Hershberg, James. *James B. Conant: Harvard to Hiroshima and the Making of the Nuclear Age.* New York: Knopf, 1993.

Hewlett, Richard G., and Oscar E. Anderson. *A History of the United States Atomic Energy Commission, Vol. I: The New World, 1939–1946.* Berkeley: University of California Press, 1962.

Hewlett, Richard G., and Jack M. Holl. *Atoms for Peace and War, 1953–1961: Eisenhower and the Atomic Energy Commission.* Berkeley: University of California Press, 1989.

Hitchcock, William. *France Restored: Cold War Diplomacy and the Quest for Leadership in Europe, 1944–1954.* Chapel Hill: University of North Carolina Press, 1998.

Hodge, Joseph Morgan, and Brett Bennett, eds. *Science and Empire: Knowledge and Networks of Science in the British Empire 1800–1970.* New York: Palgrave Macmillan, 2011.

Hoffmann, Dieter, and Mark Walker. *The German Physical Society in the Third Reich: Physicists between Autonomy and Accommodation.* Cambridge: Cambridge University Press, 2012.

Holloway, David. *Stalin and the Bomb: The Soviet Union and Atomic Energy, 1939–56.* New Haven, CT: Yale University Press, 1994.

Houghton, Bernard. *Out of the Dinosaurs: The Evolution of the National Lending Library for Science and Technology.* London: Clive Bingley, 1972.

Hughes, Thomas. *American Genesis: A Century of Invention and Technological Enthusiasm, 1870–1970.* 2nd ed. New York: Viking, 2004.

Hunt, Linda. *Secret Agenda: The United States Government, Nazi Scientists, and Project Paperclip, 1945 to 1990.* New York: St. Martin's Press, 1991.

International Cooperation Administration. *European Productivity and Technical Assistance Programs: A Summing Up (1948–58).* Paris: International Cooperation Administration, Technical Cooperation Division, USRO, 1958.

Jacobsen, Annie. *Operation Paperclip: The Secret Intelligence Program That Brought Nazi Scientists to America.* New York: Little, Brown, 2014.

Jasanoff, Sheila. *Designs on Nature: Science and Democracy in Europe and the United States.* Princeton, NJ: Princeton University Press, 2007.

Jay, Martin. *Permanent Exiles: Essays on the Intellectual Migration from Germany to America.* New York: Columbia University Press, 1985.

Jewett, Andrew. *Science, Democracy, and the American University: From the Civil War to the Cold War.* Cambridge: Cambridge University Press, 2012.

Johnson, Franklyn A. *Defence by Ministry: British Ministry of Defence, 1944–1974.* Teaneck, NJ: Holmes and Meier Publishers, 1980.

Jones, R. V. *Most Secret War.* London: Hamilton, 1978.

———. *The Wizard War: British Scientific Intelligence, 1939–1945.* New York: Coward, McCann and Geoghegan, 1978.

Josephson, Paul R. *Red Atom: Russia's Nuclear Power Program from Stalin to Today.* New York: W. H. Freeman, 2000.

Judt, Matthias, and Burghard Ciesla, eds. *Technology Transfer Out of Germany after 1945*. Amsterdam: Harwood Academic Publishers, 1996.

Junker, Detlef, ed. *The United States and Germany in the Era of the Cold War, 1945–1968: A Handbook, Volume 1*. Cambridge: Cambridge University Press, 2004.

Kaiser, David. *Drawing Theories Apart: The Dispersion of Feynman Diagrams in Postwar Physics*. Chicago: University of Chicago Press, 2009.

Karlsch, Rainer. *Allein Bezahlt? Die Reparationsleistungen der SBZ/DDR 1945–53*. Berlin: Ch. Links Verlag, 2013.

Karlsch, Rainer, Jochen Laufer, and Friederike Sattler. *Sowjetische Demontagen in Deutschland 1944–1949: Hintergründe, Ziele und Wirkungen*. Berlin: Duncker and Humblot, 2002.

Kevles, Daniel J. *The Physicists: The History of a Scientific Community in Modern America, Revised Edition*. New York: Knopf, 1995.

Khan, B. Zorina. *The Democratization of Invention: Patents and Copyrights in American Economic Development, 1790–1920*. Cambridge: Cambridge University Press, 2005.

Klehr, Harvey, John Earl Haynes, and Kirill Mihailovic Anderson. *The Soviet World of American Communism*. New Haven, CT: Yale University Press, 1998.

Kohler, Robert E. *Partners in Science: Foundations and Natural Scientists, 1900–1945*. Chicago: University of Chicago Press, 1991.

Kohlrausch, Martin, and Helmuth Trischler. *Building Europe on Expertise: Innovators, Organizers, Networkers*. New York: Palgrave Macmillan, 2014.

Kojevnikov, Alexei B. *Stalin's Great Science: The Times and Adventures of Soviet Physicists*. London: Imperial College Press, 2006.

Kolodziej, Edward A. *Making and Marketing Arms: The French Experience and Its Implications for the International System*. Princeton, NJ: Princeton University Press, 2014.

Krementsov, Nikolai. *Stalinist Science*. Princeton, NJ: Princeton University Press, 1997.

Krige, John. *American Hegemony and the Postwar Reconstruction of Science in Europe*. Cambridge, MA: MIT Press, 2006.

Krige, John, Angelina Long Callahan, and Ashok Maharaj. *NASA in the World: Fifty Years of International Collaboration in Space*. New York: Palgrave Macmillan, 2013.

Krohn, Claus-Dieter. *Intellectuals in Exile: Refugee Scholars and the New School for Social Research*. Amherst: University of Massachusetts Press, 1993.

Kurowski, Franz. *Alliierte Jagd Auf Deutsche Wissenschafler: Das Unternehmen Paperclip*. München: Kristall bei Langen Müller, 1982.

Laney, Monique. *German Rocketeers in the Heart of Dixie: Making Sense of the Nazi Past during the Civil Rights Era*. New Haven, CT: Yale University Press, 2015.

Lasby, Clarence G. *Project Paperclip: German Scientists and the Cold War*. New York: Atheneum, 1971.

Latham, Michael E. *The Right Kind of Revolution: Modernization, Development, and US Foreign Policy from the Cold War to the Present*. Ithaca, NY: Cornell University Press, 2011.

Lehrer, Tom. *Wernher von Braun*. Audio CD. Vol. The Tom Lehrer Collection. Shout Factory, 2010.

Leonard, Nelson J. *More Than a Memoir*. Philadelphia: Xlibris, 2006.

Lesch, John, ed. *The German Chemical Industry in the Twentieth Century*. Dordrecht: Kluwer, 2000.

Leslie, Stuart W. *The Cold War and American Science: The Military-Industrial-Academic Complex at MIT and Stanford*. New York: Columbia University Press, 1993.

Lewis, John Wilson, and Litai Xue. *China Builds the Bomb*. Stanford, CA: Stanford University Press, 1988.

Lind, Michael. *Land of Promise: An Economic History of the United States*. New York: Harper Paperbacks, 2013.

Lobel, Orly. *Talent Wants to Be Free: Why We Should Learn to Love Leaks, Raids, and Free Riding.* New Haven, CT: Yale University Press, 2013.

Longden, Sean. *T-Force: The Race for Nazi War Secrets, 1945.* London: Constable, 2010.

MacIsaac, David. *Strategic Bombing in World War II: The Story of the United States Strategic Bombing Survey.* New York: Garland, 1976.

MacLeod, Christine. *Inventing the Industrial Revolution: The English Patent System, 1660–1800.* Cambridge: Cambridge University Press, 1988.

Macrakis, Kristie. *Seduced by Secrets: Inside the Stasi's Spy-Tech World.* Cambridge: Cambridge University Press, 2006.

———. *Surviving the Swastika: Scientific Research in Nazi Germany.* Oxford: Oxford University Press, 1993.

Maddrell, Paul. *Spying on Science: Western Intelligence in Divided Germany, 1945–1961.* Oxford: Oxford University Press, 2008.

Maier, Helmut. *Forschung als Waffe: Rüstungsforschung in der Kaiser-Wilhelm-Gesellschaft und das Kaiser-Wilhelm-Institut für Metallforschung, 1900–1945/48.* Göttingen: Wallstein, 2007.

Manchester, William. *The Arms of Krupp, 1587–1968: The Rise and Fall of the Industrial Dynasty That Armed Germany at War.* Boston: Back Bay Books, 2003.

Margolian, Howard. *Unauthorized Entry: The Truth about Nazi War Criminals in Canada, 1946–1956.* Toronto: University of Toronto Press, 2000.

Mastanduno, Michael. *Economic Containment: CoCom and the Politics of East-West Trade.* Ithaca, NY: Cornell University Press, 1992.

May, Christopher, and Susan K. Sell. *Intellectual Property Rights: A Critical History.* Boulder, CO: Lynne Rienner Publishers, 2006.

Meckler, Alan M. *Micropublishing: A History of Scholarly Micropublishing in America, 1938–1976.* Westport, CT: Greenwood Press, 1982.

Meehan, Patricia. *A Strange Enemy People: Germans under the British, 1945–50.* London: Peter Owen, 2001.

Mick, Christoph. *Forschen für Stalin: Deutsche Fachleute in der sowjetischen Rüstungsindustrie 1945–1958.* München: Oldenbourg, 2000.

Mikhalov, Aleksandr Ivanovič, A. I. Chernyi, Rudžero Sergeevič Giljarevskij, and Robert H. Burger. *Scientific Communications and Informatics.* Arlington, VA: Information Resources Press, 1984.

Milward, Alan S. *The Reconstruction of Western Europe, 1945–51.* Berkeley: University of California Press, 1984.

Modelski, George A. *Atomic Energy in the Communist Bloc.* Melbourne: Melbourne University Press, 1959.

Naimark, Norman M. *The Russians in Germany: A History of the Soviet Zone of Occupation, 1945–1949.* Cambridge, MA: The Belknap Press of Harvard University Press, 1995.

Neufeld, Michael J. *The Rocket and the Reich: Peenemünde and the Coming of the Ballistic Missile Era.* New York: Free Press, 1995.

———. *Von Braun: Dreamer of Space, Engineer of War.* New York: Knopf, 2007.

Oberkrome, Willi, and Karin Orth. *Die Deutsche Forschungsgemeinschaft 1920–1970: Forschungsförderung im Spannungsfeld von Wissenschaft und Politik.* Stuttgart: Steiner, 2010.

Palmer, Scott W. *Dictatorship of the Air: Aviation Culture and the Fate of Modern Russia.* Cambridge: Cambridge University Press, 2006.

Parshall, Karen, and Adrian C. Rice. *Mathematics Unbound: The Evolution of an International Mathematical Research Community, 1800–1945.* Providence, RI: American Mathematical Society, 2002.

Pash, Boris. *The Alsos Mission.* New York: Award House, 1969.

Patentamt, Deutsches. *Hundert Jahre Patentamt.* München: Heymann, 1977.

Peltzer, Lilli. *Die Demontage Deutscher Naturwissenschaftlicher Intelligenz Nach Dem 2. Weltkrieg.* Berlin: ERS-Verlag, 1995.

Polanyi, Michael. *Personal Knowledge: Towards a Post-Critical Philosophy.* Chicago: University of Chicago Press, 1958.

———. *The Tacit Dimension.* Chicago: University of Chicago Press, 1966.

Pollock, Ethan. *Stalin and the Soviet Science Wars.* Princeton, NJ: Princeton University Press, 2006.

Pottage, Alain, and Brad Sherman. *Figures of Invention: A History of Modern Patent Law.* Oxford: Oxford University Press, 2010.

Price, Derek J. de Solla. *Little Science, Big Science.* New York: Columbia University Press, 1963.

———. *Science since Babylon.* New Haven, CT: Yale University Press, 1961.

Renneberg, Monika, and Mark Walker. *Science, Technology, and National Socialism.* Cambridge: Cambridge University Press, 1994.

Rhodes, Richard. *The Making of the Atomic Bomb.* New York: Simon and Schuster, 1986.

Richards, Pamela Spence. *Scientific Information in Wartime: The Allied-German Rivalry, 1939–1945.* Westport, CT: Greenwood Press, 1994.

Richelson, Jeffrey. *Spying on the Bomb: American Nuclear Intelligence from Nazi Germany to Iran and North Korea.* New York: W. W. Norton, 2007.

Rieger, Bernhard. *Technology and the Culture of Modernity in Britain and Germany, 1890–1945.* Cambridge: Cambridge University Press, 2005.

Riehl, Nikolaus. *Zehn Jahre im Goldenen Käfig: Erlebnisse Beim Aufbau der Sowjetischen URAN-Industrie.* Stuttgart: Riederer, 1988.

Riehl, Nikolaus, and Frederick Seitz. *Stalin's Captive: Nikolaus Riehl and the Soviet Race for the Bomb.* Washington, DC: American Chemical Society, 1996.

Rivas, Darlene. *Missionary Capitalist: Nelson Rockefeller in Venezuela.* Chapel Hill: University of North Carolina Press, 2001.

Rollings, Neil. *British Business in the Formative Years of European Integration, 1945–1973.* Cambridge: Cambridge University Press, 2007.

Sherman, Brad, and Lionel Bently. *The Making of Modern Intellectual Property Law.* Cambridge: Cambridge University Press, 2008.

Sibley, Katherine A. *Red Spies in America: Stolen Secrets and the Dawn of the Cold War.* Lawrence: University Press of Kansas, 2007.

Siddiqi, Asif A. *The Red Rockets' Glare: Spaceflight and the Soviet Imagination, 1857–1957.* Cambridge: Cambridge University Press, 2014.

———. *Sputnik and the Soviet Space Challenge, Vol. 1.* Gainesville: University Press of Florida, 2003.

———. *Sputnik and the Soviet Space Challenge, Vol. 2.* Gainesville: University Press of Florida, 2003.

Siegmund-Schultze, Reinhard. *Mathematicians Fleeing from Nazi Germany.* Princeton, NJ: Princeton University Press, 2009.

Simpson, Christopher. *Blowback: America's Recruitment of Nazis and Its Effects on the Cold War.* New York: Weidenfeld and Nicolson, 1988.

Slaveski, Filip. *The Soviet Occupation of Germany: Hunger, Mass Violence and the Struggle for Peace, 1945–1947.* Cambridge: Cambridge University Press, 2013.

Slusser, Robert M., ed. *Soviet Economic Policy in Postwar Germany: A Collection of Papers by Former Soviet Officials.* New York: Research Program on the U.S.S.R., 1953.

Sokolov, Valentin L. *Soviet Use of German Science and Technology, 1945–1946.* New York: Research Program on the U.S.S.R., 1955.

Solovey, Mark. *Shaky Foundations: The Politics-Patronage-Social Science Nexus in Cold War America.* New Brunswick, NJ: Rutgers University Press, 2013.

Spitz, Peter. *Petrochemicals: The Rise of an Industry*. New York: Wiley, 1988.

Steen, Kathryn. *The American Synthetic Organic Chemicals Industry: War and Politics, 1910–1930*. Chapel Hill: University of North Carolina Press, 2014.

Stent, Gunther S. *Nazis, Women, and Molecular Biology: Memoirs of a Lucky Self-Hater*. Kensington, CA: Briones Books, 1998.

Stimson, Henry L., and McGeorge Bundy. *On Active Service in Peace and War*. New York: Harper and Brothers, 1948.

Stokes, Raymond. *Constructing Socialism: Technology and Change in East Germany 1945–1990*. Baltimore: Johns Hopkins University Press, 2000.

———. *Divide and Prosper: The Heirs of I. G. Farben under Allied Authority, 1945–1951*. Berkeley: University of California Press, 1988.

———. *Opting for Oil: The Political Economy of Technological Change in the West German Chemical Industry, 1945–1961*. Cambridge: Cambridge University Press, 1994.

Stuart, Douglas. *Creating the National Security State: A History of the Law That Transformed America*. Princeton, NJ: Princeton University Press, 2012.

Sutton, Anthony C. *Western Technology and Soviet Economic Development, 1945 to 1965*. Stanford, CA: Hoover Institution Press, 1973.

Tarbell, D. *Roger Adams: Scientist and Statesman*. Washington, DC: American Chemical Society, 1981.

Tooze, Adam. *The Wages of Destruction: The Making and Breaking of the Nazi Economy*. New York: Penguin, 2006.

Turner, Ian D., ed. *Reconstruction in Post-War Germany: British Occupation Policy and the Western Zones, 1945–55*. New York: St. Martin's Press, 1989.

Uhl, Matthias. *Stalins V-2: Der Technologietransfer der Deutschen Fernlenkwaffentechnik in die UdSSR und der Aufbau der Sowjetischen Raketenindustrie 1945 bis 1959*. Bonn: Bernard and Graefe, 2001.

Vierhaus, Rudolf, and Bernhard vom Brocke, eds. *Forschung im Spannungsfeld von Politik und Gesellschaft: Geschichte und Struktur der Kaiser-Wilhelm-/Max-Planck-Gesellschaft: Aus Anlass ihres 75jährigen Bestehens*. Stuttgart: Deutsche Verlags-Anstalt, 1990.

vom Brocke, Bernhard, and Hubert Laitko, eds. *Die Kaiser-Wilhelm-/Max-Planck-Gesellschaft und ihre Institute: Studien zu ihrer Geschichte*. Berlin: De Gruyter, 1996.

Walker, Mark. *German National Socialism and the Quest for Nuclear Power, 1939–49*. Cambridge: Cambridge University Press, 1990.

———. *Nazi Science: Myth, Truth, and the German Atomic Bomb*. Cambridge, MA: Perseus Publishing, 1995.

Wang, Jessica. *American Science in an Age of Anxiety: Scientists, Anticommunism, and the Cold War*. Chapel Hill: University of North Carolina Press, 1999.

Wang, Zuoyue. *In Sputnik's Shadow: The President's Science Advisory Committee and Cold War America*. New Brunswick, NJ: Rutgers University Press, 2009.

Weinstein, Allen, and Alexander Vassiliev. *The Haunted Wood: Soviet Espionage in America—The Stalin Era*. New York: Random House, 1999.

Wells, Wyatt. *Antitrust and the Formation of the Postwar World*. New York: Columbia University Press, 2002.

Werner, Anja. *The Transatlantic World of Higher Education: Americans at German Universities, 1776–1914*. New York: Berghahn Books, 2013.

Wolfe, Audra J. *Competing with the Soviets: Science, Technology, and the State in Cold War America*. Baltimore: Johns Hopkins University Press, 2013.

Wright, Alex. *Cataloging the World: Paul Otlet and the Birth of the Information Age*. Oxford: Oxford University Press, 2014.

Yergin, Daniel. *Shattered Peace: The Origins of the Cold War and the National Security State.* Boston: Houghton Mifflin, 1977.

Zeitlin, Jonathan, and Gary Herrigel, eds. *Americanization and Its Limits: Reworking US Technology and Management in Post-War Europe and Japan.* Oxford: Oxford University Press, 2000.

Zimmermann, Paul A. *Patentwesen in der Chemie: Urspruenge, Anfaenge, Entwicklung.* Ludwigshafen: BASF Aktiengesellschaft, 1965.

Articles and Book Chapters

Aldrich, Richard J. "British Intelligence and the Anglo-American 'Special Relationship' during the Cold War." *Review of International Studies* 24, no. 3 (1998): 331–51.

Aspray, William. "Command and Control, Documentation, and Library Science: The Origins of Information Science at the University of Pittsburgh." *IEEE Annals of the History of Computing* 21, no. 4 (1999): 4–20.

Augustine, Dolores. "Wunderwaffen of a Different Kind: Nazi Scientists in East German Industrial Research." *German Studies Review* 29, no. 3 (2006): 579–88.

Beach, Robert E. "A Question of Property Rights: The Government and Industrial Know-How." *A.B.A. Journal* 41, no. 11 (1955): 1024–7.

Behrman, J. N. "Licensing Abroad under Patents, Trademarks, and Know-How by US Companies." *Patent, Trademark and Copyright Journal of Research and Education* 2 (1958): 181–277.

Behrman, J. N., and W. E. Schmidt. "New Data on Foreign Licensing." *Patent, Trademark and Copyright Journal of Research and Education* 3 (1959): 357–88.

Berghahn, Volker. "The Debate on 'Americanization' among Economic and Cultural Historians." *Cold War History* 10, no. 1 (February 2010): 107–30.

———. "West German Reconstruction and American Industrial Culture, 1945–1960." In *The American Impact on Postwar Germany*, edited by Reiner Pommerin, 65–81. Providence, RI: Berghahn Books, 1997.

Beyler, Richard H. "Hostile Environmental Intellectuals? Critiques and Counter-Critiques of Science and Technology in West Germany after 1945." *Berichte zur Wissenschaftsgeschichte* 31, no. 4 (December 1, 2008): 393–406.

Biagioli, Mario. "Science, Modernity, and the 'Final Solution.'" In *Probing the Limits of Representation: Nazism and the "Final Solution,"* edited by Saul Friedlander, 185–205. Cambridge, MA: Harvard University Press, 1992.

Bleeke, Joel A., and James A. Rahl. "The Value of Territorial and Field-of-Use Restrictions in the International Licensing of Unpatented Know-How: An Empirical Study." *Northwest Journal of International Law and Business* 1, no. 2 (1979): 450–91.

Boncourt, Thibaud. "L'internationalisation de la Science Politique: Une Comparaison Franco-Britannique (1945–2010)." Université Montesquieu–Bordeaux IV, 2011. Accessed June 28, 2016. http://tel.archives-ouvertes.fr/tel-00606354.

Bossuat, M. Gérard. "Armements et Relations Franco-Allemandes (1945–1963): Coopération et Recherche de Puissance." In *Histoire de l'Armement en France (1914–1962)*, Centre d'Histoire et d'Études des Armements, 147–206. Paris: Addim, 1995.

Bourquin, Jean-Christophe. "Le Comité National de la Recherche Scientifique: Sociologie et Histoire (1950–1967)." *Cahiers pour l'Histoire du CNRS 1939–1989* 3 (1989): 127–60.

Busquin, Phillipe. "Le Rôle des Réseaux de Scientifiques dans l'émergence d'un Espace Européen de la Recherche." In *Réseaux Économiques et Construction Européenne: Economic Networks and European Integration*, edited by Michel Dumoulin, 117–22. Brussels: PIE Peter Lang, 2004.

Carew, Anthony. "The Anglo-American Council on Productivity (1948–52): The Ideological

Roots of the Post-War Debate on Productivity in Britain." *Journal of Contemporary History* 26, no. 1 (January 1, 1991): 49–69.

Cassidy, David. "Controlling German Science, I: US and Allied Forces in Germany, 1945–1947." *Historical Studies in the Physical and Biological Sciences* 24, no. 2 (1994): 197–235.

Charpentier-Morize, Micheline. "La Contribution des 'laboratoires Propres' du CNRS à la Recherche Chimique en France de 1939 à 1973." *Cahiers pour l'Histoire du CNRS 1939–1989* 4 (1989): 79–112.

Correa, Carlos M. "Legal Nature and Contractual Conditions in Know-How Transactions." *Georgia Journal of International and Comparative Law* 11, no. 3 (1981): 450–94.

Cozzens, Susan. "The Discovery of Growth: Statistical Glimpses of Twentieth-Century Science." In *Companion to Science in the Twentieth Century*, edited by John Krige and Dominique Pestre, 127–42. New York: Routledge, 2003.

Dauchelle, Sandrine. "Le Réarmement Français Après la Seconde Guerre Mondiale." PhD dissertation, Université Paris-Sorbonne, 2006.

Davis, John R. "Friedrich Max Müller and the Migration of German Academics to Britain in the Nineteenth Century." In *Migration and Transfer from Germany to Britain, 1660–1914*, edited by Stefan Manz, Margrit Schulte Beerbühl, and John R. Davis, 93–106. München: Saur, 2007.

Day, C. R. "Science, Applied Science and Higher Education in France 1870–1945, an Historiographical Survey since the 1950s." *Journal of Social History* 26, no. 2 (1992): 367–84.

Defrance, Corine. "La mission du CNRS en Allemagne (1945–1950)." *La revue pour l'histoire du CNRS* 5 (2001). http://histoire-cnrs.revues.org/3372.

Doel, Ronald. "Roger Adams: Linking University Science with Policy on the World Stage." In *No Boundaries: University of Illinois Vignettes*, edited by Lillian Hoddeson, 124–44. Champaign: University of Illinois Press, 2004.

———. "Scientists as Policymakers, Advisors, and Intelligence Agents: Linking Diplomatic History with the History of Science." In *The Historiography of the History of Contemporary Science, Technology, and Medicine*, edited by T. Söderqvist, 33–62. Amsterdam: Harwood Academic Publishers, 1997.

Doel, Ronald, and Kristine C. Harper. "Prometheus Unleashed: Science as a Diplomatic Weapon in the Lyndon B. Johnson Administration." *Osiris* 21, no. 1 (2006): 66–85.

Dosso, Diane. "Louis Rapkine (1904–1948) et la Mobilisation Scientifique de la France Libre." PhD dissertation, University of Paris, 1998.

Eck, Jean-François. "Les Contacts Entre Groupes de l'industrie Chimique Français et Allemands de 1945 à La Fin Des Années 1960: Entre Compétition et Coopération." In *Réseaux Économiques et Construction Européenne: Economic Networks and European Integration*, edited by Michel Dumoulin, 217–34. Brussels: PIE Peter Lang, 2004.

Eckert, Michael. "Strategic Internationalism and the Transfer of Technical Knowledge: The United States, Germany, and Aerodynamics after World War I." *Technology and Culture* 46, no. 1 (2005): 104–31.

Erker, Paul. "The Long Shadow of Americanization: The German Rubber Industry and the Radial Tyre Revolution." In *Americanization and Its Limits: Reworking US Technology and Management in Post-War Europe and Japan*, edited by Jonathan Zeitlin and Gary Herrigel, 298–315. Oxford: Oxford University Press, 2000.

Ernst, Maurice C. "Economic Intelligence in CIA." In *Inside CIA's Private World: Declassified Articles from the Agency's Internal Journal, 1955–1992*, edited by H. Bradford Westerfield, 305–31. New Haven, CT: Yale University Press, 1995.

Farquharson, J. "Governed or Exploited? The British Acquisition of German Technology, 1945–48." *Journal of Contemporary History* 32, no. 1 (1997): 23–42.

Freeze, Karen Johnson. "Innovation and Technology Transfer during the Cold War: The Case of the Open-End Spinning Machine from Communist Czechoslovakia." *Technology and Culture* 48, no. 2 (2007): 249–85.

Galison, Peter. "Removing Knowledge." *Critical Inquiry* 31, no. 1 (2004): 229–43.

Gentile, Gian Peri. "Advocacy or Assessment? The United States Strategic Bombing Survey of Germany and Japan." *Pacific Historical Review* 66, no. 1 (1997): 53–79.

Germuska, Pál. "Conflicts of Eastern and Western Technology Transfer: Licenses, Espionage, and R&D in the Hungarian Defense Industry during the 1970s and 1980s." *Comparative Technology Transfer and Society* 7, no. 1 (June 18, 2009): 43–65.

Geyer, Michael. "Industriepolitik in her DDR: Von der grossindustriellen Nostalgie zum Zusammenbruch." In *Die DDR als Geschichte: Fragen, Hypothesen, Perspektiven*, edited by Jürgen Kocka and Martin Sabrow, 122–34. Berlin: Akademie-Verlag, 1994.

Giliarevskii, Ruggero. "Soviet Scientific and Technical Information System: Its Principles, Development, Accomplishments, and Defects." In *Proceedings of the 1998 Conference on the History and Heritage of Science Information Systems*, edited by Mary Ellen Bowden, Trudi Bellardo Hahn, and Robert V. Williams, 195–205. Medford, NJ: Information Today, 1999.

Gimbel, John. "German Scientists, United States Denazification Policy, and the 'Paperclip Conspiracy.'" *The International History Review* 12, no. 3 (1990): 441–65.

Goekjian, Samuel V. "Legal Problems of Transferring Technology to the Third World." *American Journal of Comparative Law* 25, no. 3 (1977): 565–70.

Greenberg, Suzanne F. "The WIPO Model Laws for the Protection of Unpatented Know-How: A Comparative Analysis." *International Tax and Business Lawyer* 3, no. 1 (1985): 52–80.

Griset, Pascal, and Yves Bouvier. "De l'histoire des techniques à l'histoire de l'innovation: Tendances de la recherche française en histoire contemporaine." *Histoire, économie and société* 31, no. 2 (2012): 29–43.

Gudmestad, Terje. "Patent Law of the United States and the United Kingdom: A Comparison." *Loyola of Los Angeles International and Comparative Law Review* 5 (1982): 173–96.

Guthleben, Denis. "La participation du Centre à l'effort scientifique de guerre." *La revue pour l'histoire du CNRS*, no. 14 (May 3, 2006): https://histoire-cnrs.revues.org/1844.

Hagen, Antje. "Patents Legislation and German FDI in the British Chemical Industry before 1914." *Business History Review* 71, no. 3 (1997): 351–80.

Hagood, Jonathan D. "Why Does Technology Transfer Fail? Two Technology Transfer Projects from Peronist Argentina." *Comparative Technology Transfer and Society* 4, no. 1 (2006): 73–98.

Hart, John D. "The ALSOS Mission, 1943–1945: A Secret U.S. Scientific Intelligence Unit." *International Journal of Intelligence and Counterintelligence* 18, no. 3 (2005): 508–37.

Heald, Lionel. "Patent Law: The British Trend." *American Bar Association Journal* 38 (1952): 217–33.

Home, R. W., and Morris F. Low. "Postwar Scientific Missions to Japan." *Isis* 84, no. 3 (1993): 527–37.

Jasanoff, Sheila. "Biotechnology and Empire: The Global Power of Seeds and Science." *Osiris* 21, no. 1 (2006): 273–92.

Kaiser, David. "The Atomic Secret in Red Hands? American Suspicions of Theoretical Physicists during the Early Cold War." *Representations* 90, no. 1 (2005): 28–60.

———. "Cold War Requisitions, Scientific Manpower, and the Production of American Physicists after World War II." *Historical Studies in the Physical and Biological Sciences* 33 (2002): 131–59.

———. "Nuclear Democracy: Political Engagement, Pedagogical Reform, and Particle Physics in Postwar America." *Isis* 93, no. 2 (2002): 229–68.

Kelly, Dominic. "The International Chamber of Commerce." *New Political Economy* 10, no. 2 (2005): 259–71.

Kevles, Daniel J. "Cold War and Hot Physics: Science, Security, and the American State, 1945–56." *Historical Studies in the Physical and Biological Sciences* 20, no. 2 (1990): 239–64.

———. "'Into Hostile Political Camps': The Reorganization of International Science in World War I." *Isis* 62, no. 1 (1971): 47–60.

———. "The National Science Foundation and the Debate over Postwar Research Policy, 1942–1945: A Political Interpretation of Science—The Endless Frontier." *Isis* 68, no. 241 (1977): 5–26.

Kiersch, Gerhard. "Die französische Deutschlandpolitik 1945–1949." In *Die Deutschlandpolitik Frankreichs und die Französische Zone, 1945–1949*, edited by Claus Scharf and Hans-Jürgen Schröder, 61–76. Wiesbaden: F. Steiner, 1983.

Kipping, Matthias. "A Slow and Difficult Process: The Americanization of the French Steel-Producing and Using Industries after the Second World War." In *Americanization and Its Limits: Reworking US Technology and Management in Post-War Europe and Japan*, edited by Jonathan Zeitlin and Gary Herrigel, 209–35. Oxford: Oxford University Press, 2000.

Kirchberger, Ulrike. "Deutsche Naturwissenschaftler im Britischen Empire: Die Erforschung der Außereuropäischen Welt im Spannungsfeld Zwischen Deutschem und Britischem Imperialismus." *Historische Zeitschrift* 271, no. 3 (2000): 621–60.

Kramer, Alan. "British Dismantling Politics, 1945–1949: A Reassessment." In *Reconstruction in Post-War Germany*, edited by Ian D. Turner, 125–54. New York: Berg, 1989.

Krammer, Arnold. "German Energy Secrets Recycled." *German Studies Review* 2, no. 2 (1979): 225–27.

———. "Technology Transfer as War Booty: The U.S. Technical Oil Mission to Europe, 1945." *Technology and Culture* 22, no. 1 (1981): 68–103.

Krige, John. "Atoms for Peace, Scientific Internationalism, and Scientific Intelligence." *Osiris* 21, no. 1 (2006): 161–81.

———. "The Peaceful Atom as Political Weapon: Euratom and American Foreign Policy in the Late 1950s." *Historical Studies in the Natural Sciences* 38, no. 1 (2008): 5–44.

———. "Technological Leadership and American Soft Power." In *Soft Power and US Foreign Policy: Theoretical, Historical and Contemporary Perspectives*, edited by Inderjeet Parmar and Michael Cox, 121–26. New York: Routledge, 2010.

Kuenzel, Christiane. "Verwaltung Sowjetische [Staatliche] Aktiengesellschaften in Deutschland (SAG)." In *SMAD-Handbuch: Die Sowjetische Militäradministration in Deutschland 1945–1949*, edited by Jan Foitzik, T. V. Tsarevskaia-Diakina, Burghard Ciesla, Horst Möller, and A. O Chubarian, 388–95. München: Oldenbourg, 2009.

Ladas, Stephen P. "Legal Protection of Know-How." *Patent, Trademark, Copyright Journal of Research and Education* 7 (1963): 397–417.

Layton, Edwin T. "Mirror-Image Twins: The Communities of Science and Technology in 19th-Century America." *Technology and Culture* 12, no. 4 (1971): 562–80.

Ludmann-Obier, Marie-France. "Un Aspect de la Chasse aux Cerveaux: Les Transferts de Techniciens Allemands, 1945–1949." *Relations internationales* 46 (1986): 196–97.

———. "La Mission du CNRS en Allemagne (1945–1950)." *Cahiers pour l'Histoire du CNRS 1939–1989* 3 (1989): 1–17.

MacDonald, Bertrum H. "Implications of Nationalism in the Development of the Scientific Information Infrastructure in North America in the Decades Leading Up to World War II." In *The History and Heritage of Scientific and Technological Information Systems: Proceedings of the 2002 Conference*, edited by W. Boyd Rayward and Mary Ellen Bowden, 215–27. Medford, NJ: Information Today, 2004.

Macdonald, David R. "Know-How Licensing and the Antitrust Laws." *Trademark Reporter* 54 (1964): 251–79.

MacLeod, R. "Secrets among Friends: The Research Information Service and the 'Special Relationship' in Allied Scientific Information and Intelligence, 1916–1918." *Minerva* 37, no. 3 (1999): 201–33.

Maddrell, Paul. "British-American Scientific Intelligence Collaboration during the Occupation of Germany." *Intelligence and National Security* 15, no. 2 (2000): 74–94.

———. "Western Intelligence Gathering and the Division of German Science." *Cold War International History Project Bulletin* 12, no. 13 (1997): 352–59.

McClay, Wilfred M. "A Haunted Legacy: The German Refugee Intellectuals and American Social Thought, 1932–1969." PhD dissertation, Johns Hopkins University, 1986.

McGlade, Jacqueline. "Americanization: Ideology or Process? The Case of the United States Technical Assistance and Productivity Programme." In *Americanization and Its Limits: Reworking US Technology and Management in Post-War Europe and Japan*, edited by Jonathan Zeitlin and Gary Herrigel, 53–75. Oxford: Oxford University Press, 2000.

Medhurst, Martin J. "Atoms for Peace and Nuclear Hegemony: The Rhetorical Structure of a Cold War Campaign." *Armed Forces and Society* 23, no. 4 (1997): 571–93.

Mehos, Donna C., and Suzanne M. Moon. "The Uses of Portability: Circulating Experts in the Technopolitics of Cold War and Decolonization." In *Entangled Geographies: Empire and Technopolitics in the Global Cold War*, edited by Gabrielle Hecht, 43–74. Cambridge, MA: MIT Press, 2012.

Meshbesher, Thomas M. "The Role of History in Comparative Patent Law." *Journal of the Patent and Trademark Office Society* 78, no. 9 (1996): 594–614.

Morris, Peter J. T. "Transatlantic Transfer of Buna S Synthetic Rubber Technology, 1932–45." In *The Transfer of International Technology: Europe, Japan, and the USA in the Twentieth Century*, edited by David J. Jeremy, 57–89. Northampton, MA: Edward Elgar, 1992.

Morton, David L. "'The Rusty Ribbon': John Herbert Orr and the Making of the Magnetic Recording Industry, 1945–1960." *Business History Review* 67, no. 4 (1993): 589–622.

Moser, Petra, Alessandra Voena, and Fabian Waldinger. "German-Jewish Emigres and U.S. Invention." *American Economic Review* 104, no. 10 (2014): 3222–55.

Muddiman, Dave. "Red Information Science: J. D. Bernal and the Nationalization of Scientific Information in Britain from 1930 to 1949." In *The History and Heritage of Scientific and Technological Information Systems: Proceedings of the 2002 Conference*, edited by W. Boyd Rayward and Mary Ellen Bowden, 258–66. Medford, NJ: Information Today, 2004.

Nash, John B. "The Concept of 'Property' in Know-How as a Growing Area of Industrial Property: Its Sale and Licensing." *Patent, Trademark, Copyright Journal of Research and Education* 6 (1962): 289–98.

Neufeld, Michael J. "The Nazi Aerospace Exodus: Towards a Global, Transnational History." *History and Technology* 28, no. 1 (2012): 49–67.

Nilsson, Mikael. "The Power of Technology: U.S. Hegemony and the Transfer of Guided Missiles to NATO during the Cold War, 1953–1962." *Comparative Technology Transfer and Society* 6, no. 2 (2008): 127–49.

O'Reagan, Douglas. "Know-How in Postwar Business and the Law." *Technology and Culture* 58, no. 1 (2017): 121–53.

Pestre, Dominique. "Guerre, Renseignement Scientifique et Reconstruction, France, Allemagne et Grande-Bretagne dans les Années 1940." In *De La Diffusion Des Sciences à l'espionnage Industriel, XVe–XXe Siècle: Actes du Colloque de Lyon (30–31 Mai 1996) de La SFHST*, edited by André Guillerme, 183–201. Fontenay-aux-Roses, France: ENS Editions, 1999.

Picard, Jean Francois. "La Longue Marche vers le CNRS (1901–1945)." *Cahiers pour l'Histoire du CNRS 1939–1989* 1 (1988): 7–40.

Presas i Puig, Albert. "Reflections on a Peripheral Paperclip Project: A Technological Innovation System in Spain Based on the Transfer of German Technology." Preprint. Max Planck Institute for the History of Science, 2008.

Prost, Antoine. "Les Origines des Politiques de la Recherche en France (1939–1958)." *Cahiers pour l'Histoire du CNRS 1939–1989* 1 (1988): 41–62.

Radice, Mark. "The Hunt for Nazi Scientists." *Secrets of the Dead*. PBS, October 19, 2005.

Radkau, Joachim. "Revoltierten die Produktivkräfte Gegen den Real Existierenden Sozialismus?" *1999* 5, no. 4 (1990): 13–42.

Rammer, Gerhard. "Die Nazifizierung und Entnazifizierung der Physik an der Universität Göttingen." PhD dissertation, Georg-August-Universität Göttingen, 2004.

Ramskjaer, Liv. "Users and Producers of Plastics in Post–World War II Norway: Building a New Industry through Transfer of Technology." *Comparative Technology Transfer and Society* 3, no. 1 (2005): 76–102.

Rayward, W. Boyd. "The Origins of Information Science and the International Institute of Bibliography / International Federation for Information and Documentation (FID)." In *Historical Studies in Information Science*, edited by Trudi Bellardo Hahn and Michael Buckland, 22–33. Medford, NJ: Information Today, 1998.

Reynolds, D. "'A Special Relationship'? America, Britain and the International Order since the Second World War." *International Affairs* 62, no. 1 (1985): 1–20.

Rhees, D. J. "The Chemists' War: The Impact of World War I on the American Chemical Profession." *Bulletin of the History of Chemistry* 13/14 (1992/1993): 40–47.

Richards, Pamela Spence. "Aslib at War: The Brief but Intrepid Career of a Library Organization as a Hub of Allied Scientific Intelligence 1942–1945." *Journal of Education for Library and Information Science* 29, no. 4 (1989): 279–96.

———. "The Movement of Scientific Knowledge from and to Germany under National Socialism." *Minerva* 28, no. 4 (1990): 401–25.

———. "Scientific Information for Stalin's Laboratories, 1945–1953." In *Historical Studies in Information Science*, edited by Trudi Bellardo Hahn and Michael Buckland, 193–204. Medford, NJ: Information Today, 1998.

———. "The Soviet Overseas Information Empire and the Implications of Its Disintegration." In *Proceedings of the 1998 Conference on the History and Heritage of Science Information Systems*, edited by Mary Ellen Bowden, Trudi Bellardo Hahn, and Robert V. Williams, 206–14. Medford, NJ: Information Today, 1999.

Richelson, Jeffrey. "When Kindness Fails: Assassination as a National Security Option." *International Journal of Intelligence and CounterIntelligence* 15, no. 2 (2002): 243–74.

Ryan, Mike H. "The Role of National Culture in the Space-Based Technology Transfer Process." *Comparative Technology Transfer and Society* 2, no. 1 (2004): 31–66.

Sachse, Carola. "'Whitewash Culture': How the Kaiser Wilhelm / Max Planck Society Dealt with the Nazi Past." In *The Kaiser Wilhelm Society under National Socialism*, edited by Susanne Helm, Carola Sachse, and Mark Walker, 373–99. Cambridge: Cambridge University Press, 2009.

Schmid, Sonja D. "Nuclear Colonization? Soviet Technopolitics in the Second World." In *Entangled Geographies: Empire and Technopolitics in the Global Cold War*, edited by Gabrielle Hecht, 125–54. Cambridge, MA: MIT Press, 2012.

Schmidhuber, Juergen. "Evolution of National Nobel Prize Shares in the 20th Century." *arXiv: 1009.2634 [physics]* (September 14, 2010). http://arxiv.org/abs/1009.2634.

Schmidt, Lajos. "Licensing Know-How, Patents, and Trademarks Abroad." *University of Illinois Law Forum* 1959 (1959): 248–70.

Schneider, Ullrich. "Britische Besatzungspolitik 1945: Besatzungsmacht, deutsche Exekutive und die Probleme der unmittelbaren Nachkriegszeit, dargestellt am Beispiel des späteren Landes Niedersachsen von April bis Oktober 1945." PhD dissertation, University of Hanover, 1980.

Seely, Bruce Edsall. "Historical Patterns in the Scholarship of Technology Transfer." *Comparative Technology Transfer and Society* 1, no. 1 (2003): 7–48.

Seely, Bruce Edsall, Donald E. Klingner, and Gary Klein. "'Push' and 'Pull' Factors in Technology Transfer: Moving American-Style Highway Engineering to Europe, 1945–1965." *Comparative Technology Transfer and Society* 2, no. 3 (2004): 229–46.

Seidel, Robert W. "Secret Scientific Communities: Classification and Scientific Communication in the DOE and DOD." In *Proceedings of the 1998 Conference on the History and Heritage of Science Information Systems*, edited by Mary Ellen Bowden, Trudi Bellardo Hahn, and Robert V. Williams, 46–60. Medford, NJ: Information Today, 1999.

Sharp, Walter R. "The Institutional Framework for Technical Assistance." *International Organization* 7, no. 3 (1953): 342–79.

Siddiqi, Asif A. "Germans in Russia: Cold War, Technology Transfer, and National Identity." *Osiris* 24, no. 1 (January 2009): 120–43.

Siegel, Irving H. "Scientific Discovery, Invention, and the Cultural Environment." *Patent, Trademark and Copyright Journal of Research and Education* 4 (1960): 233–48.

Stedman, John C. "Legal Problems in the International and Domestic Licensing of Know-How." *A.B.A. Antitrust Section* 29 (1965): 247–54.

Steen, Kathryn. "Patents, Patriotism, and 'Skilled in the Art': USA v. the Chemical Foundation, Inc., 1923–1926." *Isis* 92, no. 1 (2001): 91–122.

———. "Technical Expertise and US Mobilization, 1917–18: High Explosives and War Gases." In *Frontline and Factory: Comparative Perspectives on the Chemical Industry at War, 1914–1924*, edited by Roy MacLeod and Jeffrey Allan Johnson, 103–22. Dordrecht: Kluwer, 2006.

Stewart, Robert K. "The Office of Technical Services: A New Deal Idea in the Cold War." *Knowledge: Creation, Diffusion, Utilization* 15, no. 1 (1993): 44–77.

Stokes, Raymond. "Gravity and the Rainbow-Makers: Some Thoughts on the Trajectory of the German Chemical Industry in the Twentieth Century." In *The German Chemical Industry in the Twentieth Century*, edited by John Lesch, 441–50. Dordrecht: Kluwer, 2000.

Stoltzenberg, Dietrich. "Scientist and Industrial Manager: Emil Fischer and Carl Duisberg." In *The German Chemical Industry in the Twentieth Century*, edited by John Lesch, 57–89. Dordrecht: Kluwer, 2000.

Till, James E. "Predecessors of Preprint Servers." *Learned Publishing* 14, no. 1 (2001): 7–13.

Tiratsoo, Nick, and Terence R. Gourvish. "'Making It Like Detroit': British Managers and American Productivity Methods, 1945–65." *Business and Economic History* 25, no. 1 (1996): 206–16.

Tiratsoo, Nick, and Jim Tomlinson. "Exporting the 'Gospel of Productivity': United States Technical Assistance and British Industry 1945–1960." *Business History Review* 71, no. 1 (March 1997): 41–81.

Tournès, Ludovic. "La Fondation Rockefeller et la Construction d'une Politique des Sciences Sociales en France (1918–1940)." *Annales* 63, no. 6 (2008): 1371–402.

Trager, Frank N., and Helen G. Trager. "Exporting and Training Experts." *Review of Politics* 24, no. 1 (1962): 88–108.

Trimble, Marketa. "The Patent System in Pre-1989 Czechoslovakia." *Scholarly Works*, no. 810 (2013). http://scholars.law.unlv.edu/facpub/810.

Truman, Harry S. "Special Message to the Congress on the Mutual Security Program." In *Public Papers of the Presidents of the United States: Harry S. Truman, 1945–1953*, 179–91. Washington, DC: Government Printing Office, 1966.

Turner, Ian D. "British Policy towards German Industry, 1945–49: Reconstruction, Restriction or Exploitation?" In *Reconstruction in Post-War Germany*, 67–92. New York: Berg, 1989.

Uttley, Matthew. "Operation 'Surgeon' and Britain's Post-War Exploitation of Nazi German Aeronautics." *Intelligence and National Security* 17, no. 2 (2002): 1–26.

Villain, Jacques. "France and the Peenemünde Legacy." In *History of Rocketry and Astronautics: Proceedings of the Twenty-Sixth History Symposium of the International Academy of Astronautics, Washington, D.C., U.S.A., 1992*, edited by Phillipe Young, 119–61. San Diego, CA: Univelt, 1997.

———. "L'apport des Scientifiques Allemandes aux Programmes de Recherche Relatifs aux Fusées et Avions à Réaction à Partir de 1945." In *La France Face aux Problemes d'armement 1945–1950*, edited by Maurice Vaïsse, 97–121. Brussels: Complexe, 1999.

Von Gizycki, Rainald. "Centre and Periphery in the International Scientific Community: Germany, France and Great Britain in the 19th Century." *Minerva* 11, no. 4 (1973): 474–94.

Wang, Zuoyue. "The Chinese Developmental State during the Cold War: The Making of the 1956 Twelve-Year Science and Technology Plan." *History and Technology* 31, no. 3 (July 3, 2015): 180–205.

Warheit, I. A. "The Library Program of the US Atomic Energy Commission Technical Information Service." *Bulletin of the Medical Library Association* 40, no. 1 (January 1952): 1–5.

Wark, W. K. "British Intelligence on the German Air Force and Aircraft Industry, 1933–1939." *Historical Journal* 25, no. 3 (1982): 627–48.

Wellerstein, Alex. "Knowledge and the Bomb: Nuclear Secrecy in the United States, 1939–2008." PhD dissertation, Harvard University, 2010.

———. "Patenting the Bomb: Nuclear Weapons, Intellectual Property, and Technological Control." *Isis* 99, no. 1 (2008): 57–87.

Westwick, Peter J. "Secret Science: A Classified Community in the National Laboratories." *Minerva* 38, no. 4 (2000): 363–91.

Wilkins, Mira. "German Chemical Firms in the United States from the Late 19th Century to Post–World War II." In *The German Chemical Industry in the Twentieth Century*, edited by John Lesch, 285–321. Dordrecht: Kluwer, 2000.

Williams, Robert V. "The Special Libraries and Documentation Movements in the United States, 1910–1960." In *Historical Studies in Information Science*, edited by Trudi Bellardo Hahn and Michael Buckland, 173–80. Medford, NJ: Information Today, 1998.

Winter, David. "The Licensing of Know-How to the Soviet Union." *Journal of World Trade Law* 1 (1967): 162–78.

Wolfe, Edward. "Restrictions in Know-How Agreements." *Antitrust Bulletin* 12 (1967): 749–72.

Zallen, Doris T. "The Rockefeller Foundation and French Research." *Cahiers pour l'Histoire du CNRS 1939–1989* 5 (1989): 35–58.

Zhang, Baichun, Jiuchun Zhang, and Fang Yao. "Technology Transfer from the Soviet Union to the People's Republic of China: 1949–1966." *Comparative Technology Transfer and Society* 4, no. 2 (2006): 105–67.